Methylotrophy and Methanogenesis

Aspects of Microbiology

Series editors

Dr. J.A. Cole, University of Birmingham

Dr. C.J. Knowles, University of Kent

Dr. D. Schlessinger, Washington University School of Medicine, USA

This series brings important topics of current interest, conceptual difficulty or controversy to the attention of second and third year undergraduates, and to more senior scientists wishing to acquaint themselves with recent developments in fields outside their own specialities. One of the principal objectives of the series is to bridge the gap between introductory texts and research literature. The series editors and the publisher have sought to achieve this objective by providing brief, inexpensive texts that will familiarize students with current research using language and illustrations carefully controlled to conform to undergraduate expectations.

1. Oral Microbiology *P. Marsh*
2. Bacterial Toxins *J. Stephen and R.A. Pietrowski*
3. The Microbial Cell Cycle *C. Edwards*
4. Bacterial Plasmids *K. Hardy*
5. Bacterial Respiration and Photosynthesis *C.W. Jones*
6. Bacterial Cell Structure *H.J. Rogers*
7. Microbial Control of Plant Pests and Diseases *J. Deacon*
8. Methylotrophy and Methanogenesis *P.J. Large*

Aspects of Microbiology 8

Methylotrophy and Methanogenesis

Peter J. Large

Senior Lecturer in Biochemistry
University of Hull

VNR **Springer Science+Business Media, LLC**

To David Peel and Rod Quayle
with affection and gratitude

Published by Van Nostrand Reinhold (UK) Co. Ltd.
Molly Millars Lane, Wokingham, Berkshire, England

© 1983 Springer Science+Business Media New York

Originally published by Van Nostrand Reinhold (UK) Co. Ltd. 1983

ISBN 978-0-442-30528-4   ISBN 978-94-009-3169-5 (eBook)
DOI 10.1007/978-94-009-3169-5

Preface

This short book attempts to give a reader who has a basic biochemical and microbiological background (one to two years at University level) an idea of the ecological, biochemical, physiological and biotechnological importance of methane, methanol and related compounds in the microbial world. Because the book covers several different scientific disciplines, readers may encounter unfamiliar terminology. The glossary at the end of the book defines the more obscure of these.

The book has been written during a period of heavy teaching commitments, and despite the helpful comments of many colleagues, it is likely that errors have crept in. As I have no co-author whom I can blame, I must accept sole responsibility for these!

I wish to thank the many friends, students and colleagues who have read all or part of the manuscript: Charlie Bamforth, Rick Gibson, Jeff Green, Theo Hansen, Wim Harder, Geoff Haywood and, above all, Hans van Dijken. Beste Hans, ik heb je met een flinke taak belast, en je hebt het graag aangenomen, zonder te klagen. Jouw ideeën en suggesties hebben een aanzienlijke verbetering in het boek tot stand gebracht. Hartelijk bedankt voor je moeite, ik stel het zeer op prijs. I would also like to thank the editors of the series, Jeff Cole and Chris Knowles, for their kind help in turning my jargon into prose. All the comments I have received have helped me to improve the book. I am grateful to Marten Veenhuis for Fig. 4.2. I would also like to thank authors who sent me useful information in advance of publication: Chris Anthony, Hans Duine, Ortwin Meyer, Rod Quayle, Hans van Dijken and Len Zatman. My grateful thanks are also due to Susan Wheeldon, who did most of the typing, and was very patient with all the alterations and corrections.

Hull, March 1982   P.J. LARGE

Abbreviations

ATP, ADP	adenosine 5'-tri- and di-phosphate, respectively
EC	Enzyme Commission number. This is a means of identifying enzymes by reference to the numbers assigned in the book *Enzyme Nomenclature 1978*, Academic Press, New York
FH_4	tetrahydrofolate
GSH	reduced form of glutathione (γ-glutamylcysteinylglycine
NAD^+, NADH	oxidized and reduced forms of nicotinamide-adenine dinucleotide, respectively
$NADP^+$, NADPH	oxidized and reduced forms of nicotinamide-adenine dinucleotide phosphate, respectively
$NAD(P)^+$, NAD(P)H	signifies that *either* NAD^+ or $NADP^+$, NADH or NADPH may be involved in a particular reaction
P_i	inorganic orthophosphate

Other abbreviations used in tables or figures are defined in the respective legends.

Names of micro-organisms. In order to avoid confusion (since a very large number of generic names in this book begin with the letter M), binomial names are not generally abbreviated. In cases where single letter abbreviations are used for generic names, it may be assumed that the genus is the same as the one mentioned immediately before. For names of organisms consisting of a generic name followed by a strain designation but no specific name, e.g. *Pseudomonas* AM1, single-letter abbreviations are not used.

Contents

1 The Biology and Ecology of Micro-organisms Metabolizing C_1 compounds

Introduction

This book sets out to describe the metabolic processes by which micro-organisms can use and produce reduced C_1 compounds. By this term we mean compounds of carbon that are more reduced than carbon dioxide (CO_2) and which contain no carbon – carbon bonds, although some contain more than one carbon atom. A list of compounds of this type is given in Table 1.1.

Since these materials (of which the most important are methane and methanol) are more reduced than CO_2, their oxidation to CO_2 can yield energy when they are metabolized by living cells. As we shall see, bacteria using reduced C_1 compounds for growth occupy a borderline position between autotrophs and heterotrophs. Whereas autotrophs using CO_2 as carbon source require an additional energy source, heterotrophs can use organic compounds as a simultaneous source of carbon and energy. The micro-organisms that can grow on reduced C_1 compounds are thus heterotrophs as far as their energy metabolism is concerned. However, they have unusual biochemical properties which distinguish them from heterotrophs growing on multicarbon compounds. The first of these unusual features is that the energy-yielding pathway of oxidation of the growth substrate to CO_2 does not involve the tricarboxylic acid cycle. Secondly, the pathways by which their growth substrate is assimilated into cell material are usually cyclic. This generalization may not hold for methanogenic bacteria (see Chapter 2).

Why are the production and utilization of reduced C_1 compounds by micro-organisms so important? The first reason is that one of the compounds—methane—is an important energy source. Apart from vast subterranean reserves of 'fossil' methane, enormous quantities of methane gas are produced daily by the biological activity of bacteria, and scientists are interested in the possibility of utilizing these biological processes as an alternative source of energy. The aim of this book is to help you to understand firstly the biochemical and physiological processes by which living organisms produce methane and secondly the processes by which other organisms are able to grow on methane and other reduced C_1 compounds.

The second reason is that the biological processes of methane turnover occur on a massive scale. This has only recently become apparent, owing to the development in the last ten years of instruments that can measure low concentrations of methane in the atmosphere. Thus methane is considerable ecological importance.

Table 1.1   Occurrence of Reduced C_1 Compounds in Nature

Compound and formula	Occurrence	Source
Methane (CH_4)	Ubiquitous where ruminants live Anaerobic lakes and paddy fields	Methanogenic bacteria
Methanol (CH_3OH)	In small quantities in the atmosphere In plant material	Photo-oxidation of methane Decomposition of lignin, hemicelluloses and pectin
Formaldehyde (HCHO)	Trace amounts only	Waste from tanneries and chemical processing plants
Formate (HCOOH)	Industrial waste	Effluent from tanning and rubber processing Product of mixed acid fermentation
Formamide ($HCONH_2$)	Industrial waste	
Cyanide (CN^-)	Industrial waste In plant material	Electroplating and metal-extraction Product of cyanogenic plants, fungi and bacteria
Carbon monoxide (CO)	Populated urban areas Decaying organic matter	Blast-furnace and synthesis gas Exhaust from road vehicles Decomposition of porphyrins
Mono-, di- and tri-methylamine (CH_3NH_2, $(CH_3)_2NH$, $(CH_3)_3N$)	Industrial waste Food industry waste	Tanning effluent Decaying fish
Trimethylamine N-oxide ($(CH_3)_3NO$)	Ubiquitous where fish and invertebrates die	Fish and invertebrate muscle and body fluid
Dimethyl sulphide, sulphoxide and sulphone [$(CH_3)_2S$, $(CH_3)_2SO$, $(CH_3)_2SO_2$]	Industrial waste Atmosphere	Bisulphite liquor from woodpulp processing Plants and marine algae

The third reason is that man is also interested in using the chemical energy of methane and its derivatives not by direct combustion but in the form of food. Methane and C_1 compounds have no food value to highly evolved organisms, but use can be made of the micro-organisms we are going to consider to create a new link in a food chain enabling vertebrates to consume methane indirectly as food. The protein and carbohydrate from these micro-organisms could thus be used indirectly or directly as additional food sources for the rapidly increasing human population of this planet. The fourth reason is the possibility of exploiting the capabilities of micro-organisms to metabolize reduced C_1 compounds to remove possible environmental contaminants, such as carbon monoxide, cyanide and methyl sulphides.

The fifth reason is the possible utilization of enzyme systems from these micro-organisms to bring about under mild conditions in the laboratory or on an industrial scale the chemical transformation of several compounds that otherwise would require expensive, potentially dangerous, and energy-consuming industrial plant (see Chapters 3 and 5).

Since most (but not all) reduced C_1 compounds contain methyl groups, organisms that can grow on reduced C_1 compounds as sole source of carbon and energy are called *methylotrophs*. The capacity to grow on such compounds is called *methylotrophy*. What is the reason for inventing a special word for a group of organisms that are after all only a specialized kind of heterotroph? The main reason is that methylotrophs occupy an intermediate position in nature between autotrophs and heterotrophs. They have special pathways of energy generation and carbon assimilation which mark them out as an individual biological group, although the borderlines between the various categories autotroph, methylotroph and heterotroph are not always clear-cut. Some methylotrophs indeed (Chapter 3) use an autotrophic carbon-assimilation pathway, and some under certain conditions may *be* autotrophs since they can oxidize inorganic compounds as well as organic compounds to obtain energy for growth. Some methylotrophs also grow well on multicarbon compounds, and are called *facultative* methylotrophs. A few methylotrophs seem to be unable to use multicarbon compounds—they are restricted to only a few reduced C_1 compounds. Such organisms are called *obligate* methylotrophs. Certain photosynthetic bacteria can also grow anaerobically in the light on reduced C_1 compounds (Chapter 3), where the C_1 compound may function simultaneously as a source of electrons and as a source of CO_2 for carbon assimilation, but not as a source of energy, which comes from light.

The organisms that produce methane in nature (see Table 1.1) are not necessarily methylotrophs, but they are a specialized group with extremely interesting properties, and an unusual habitat and physiology (see Chapter 2).

Occurrence of Reduced C_1 Compounds in Nature

Some C_1 compounds, because they are biological products, are relatively ubiquitous in nature, e.g. methane, whereas others occur only in very limited environments, in some cases mainly as a result of man's activity, and are sometimes pollutants. An example of this group is carbon monoxide. The occurrence of C_1 compounds in nature is summarized in Table 1.1.

What is the fate of these compounds in the biosphere? They are in fact all decomposed: none is so inert that it accumulates in the environment. Although some compounds disappear as a result of chemical processes such as photo-oxidation in the upper atmosphere, most are removed from the environment by the metabolic activities of living organisms (even toxic materials like cyanide). In the case of gaseous materials like methane and carbon monoxide, it is difficult to estimate quantitatively the separate contributions of these two routes, but huge amounts of methane are thought to be photo-oxidized to methanol, formaldehyde and carbon monoxide in the upper atmosphere. In most cases the decomposition of reduced C_1 compounds by living organisms (mainly micro-organisms) also requires oxygen.

If conditions are such that aerobic micro-organisms can flourish, pollution is not usually a problem. Pollution mainly occurs when there is an excess of organic material under oxygen-deficient conditions. All dissolved oxygen is then immediately removed by microbial populations for their metabolic processes so that growth is oxygen-limited. We shall return to the problems of pollution in Chapter 5.

The Carbon Cycle

When the compounds that we are considering are broken down, they form part of the carbon cycle. All the carbon on the earth has at some time or other formed part of this cycle. If at one particular moment in history conditions are such that carbon in organic or inorganic form can be concentrated under anaerobic conditions, then that carbon ceases to circulate and becomes fossilized. Examples of fossilized carbon are limestone rocks, coal, natural gas and oil deposits. If the carbon in these is liberated by the hand of man it will return to the circulating carbon of the cycle.

This book discusses a relatively restricted group of carbon compounds, but a group nevertheless which, because it contains methane, is of a fundamental quantitative importance. The production of methane takes place only under anaerobic conditions. It is formed by strictly anaerobic bacteria that can oxidize certain compounds using CO_2 as the terminal electron acceptor. These bacteria are a specialized (and probably very ancient) group of micro-organisms that require no oxygen for growth and live in habitats that may have been unchanged for millions of years (see Chapter 2). They are called *methanogenic* (methane-producing) bacteria. They produce a staggering quantity of methane (see Table 1.2): some of it is formed by free-living methanogenic bacteria in lakes and paddy fields, and some by bacteria living in a specialized environment—the rumen or fore-stomach of cows, sheep, goats and horses (ruminant animals). The rumen methanogens form part of a symbiotic population that allows the ruminant to obtain energy from cellulose, and these organisms produce world-wide about 4×10^5 tonnes of methane per day (1 tonne $= 10^6$ g).

One of the consequences of the ubiquitous production of methane is the evolution of micro-organisms that can grow aerobically on it as sole carbon and energy source. These methylotrophs have been given the name *methano-*

trophs to signify that they can grow on methane itself. They play an important role in utilizing the oxygen of the earth's atmosphere to recycle methane into organic compounds and to make it available, as CO_2, to autotrophs. Thus, as a specialized part of the normal carbon cycle, we have the *methane cycle* (Fig. 1.1).

Carbon monoxide occurs in the atmosphere mainly as a result of human activity (emissions from blast furnaces, chemical plants and automobile exhausts). Total production has been estimated at 7×10^8 tonnes per year, and it is a serious pollutant in many urban areas. It can be oxidized by many bacteria, and some methylotrophs can use it for growth (see Chapter 3).

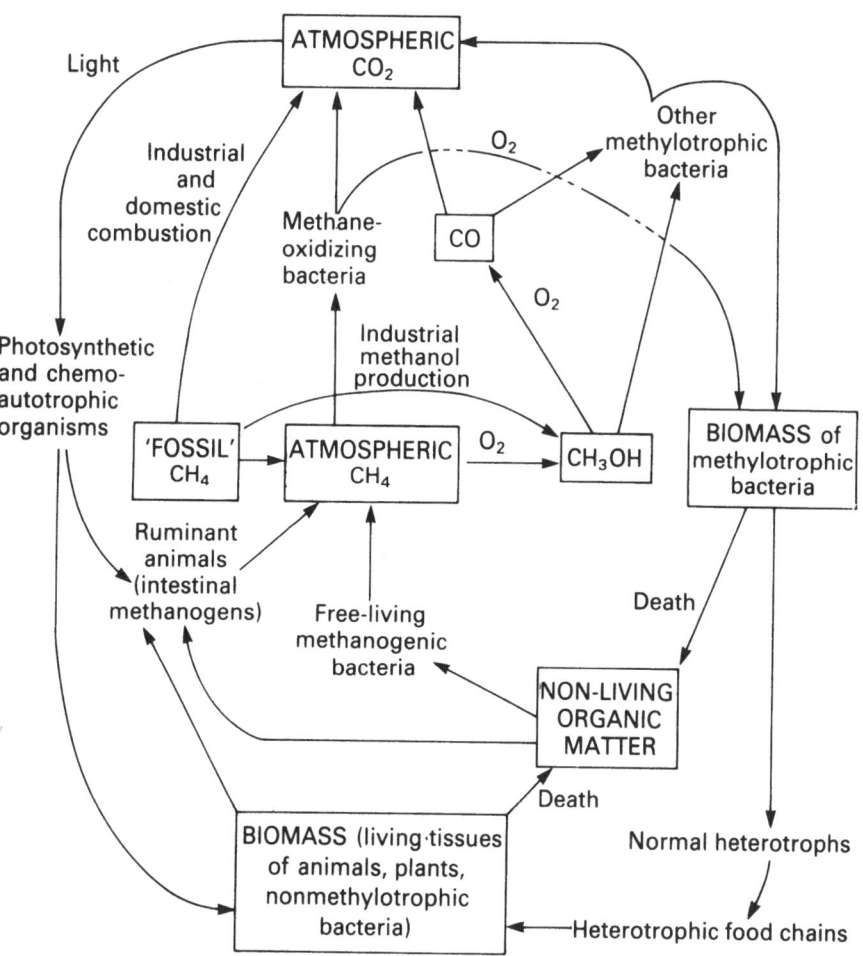

Fig. 1.1 The methane cycle.

Ecology of Methane Formation and Oxidation in Nature

There are three basic types of habitat that involve the production (and in some habitats also the consumption) of methane. These three types have been classified by Wolfe and Higgins (1979) as follows (see Table 1.2). Type A habitats are aquatic sediments, swamps, tundra, decaying heartwood of trees and the anaerobic sludge digester (Chapter 5). Type B are the gastro-intestinal habitats (the rumen, caecum and intestine), and type C habitats include geothermal springs and the African lake Kivu. In all habitats the population of methanogens is a mixture of various species, but only mesophilic and thermophilic strains can live in habitat C. The *biochemical*

Table 1.2   Sources of methane

(Data from D.H. Ehhalt (1976). The atmospheric cycle of methane. In *Symposium on Microbial Production and Utilization of Gases* (H_2, CH_4, CO). Edited by H.G. Schlegel, G. Gottschalk and N. Pfennig. By permission of Akademie der Wissenschaften, Göttingen.)

Methane in the atmosphere

Habitat (see text)		$10^{-6} \times$ Amount produced (tonnes CH_4 year^{-1})
	Biogenic sources	
B	Enteric fermentation of ruminants	101–220
A	Paddy fields	280
A	Swamps and marshes	130–260
A	Fresh-water lakes	1.25–25
A,C	Other	5.27–16.5
	Total	528–812
	Anthropogenic sources	
	Coal mining	6.3–22
	Lignite mining	1.6–5.7
	Industrial losses	7–21
	Automobile exhaust	0.5
	Volcanic	0.2
	Total	15.6–49.4

Methane used for combustion

Year	Total world output from natural gas wells (tonnes)
1965	520
1974	900

6

processes and the organisms involved in methane production are discussed in Chapter 2, but we can first consider one example of a type A habitat in which methanogenic and methanotropic bacteria coexist, namely a stratified eutrophic fresh-water lake.

Lakes rich in organic material are said to be *eutrophic*. Unless they are very shallow, lakes rarely freeze completely because the maximum density of water is at 4°C, so that the colder water floats to the top. In spring when the water warms up, only the upper layers get warm, and the deeper water tends to remain at 4°C, forming a cold layer called the *hypolimnion*. The transition between the cold layer and the warmer, lighter layer where there are convection currents—the *epilimnion*—is usually quite sharp. This sharp temperature change, the *thermocline*, varies in depth with the lake, but is often about 15 m down from the surface. If the area is very windy, the thermocline is less sharp and forms an intermediate zone or *metalimnion*. Only in spring and autumn, when the epilimnion is also at 4°C, can the two layers mix. This means that the oxygen supplied to the hypolimnion at this time of mixing has to last half a year till the next mixing season, because at 15 m depth there is no light to support photosynthetic organisms which could produce oxygen in this region (see Fig. 1.2).

In such a lake the warm, light epilimnion is a habitat for photosynthetic phytoplankton that fix carbon dioxide. The organic material so formed can be metabolized by other micro-organisms, but sooner or later organic material finds itself in the hypolimnion and eventually falls into the mud at the bottom of the lake. At the bottom of the hypolimnion, conditions are always anaerobic, and organic material is slowly metabolized by anaerobic bacteria that use sulphate or carbon dioxide as their terminal electron acceptors, producing hydrogen sulphide and methane, which bubble slowly up to the epilimnion. On or near the thermocline, where dissolved oxygen can penetrate, methane-oxidizing bacteria are established. These methanotrophs grow on methane, producing carbon dioxide and organic carbon. The latter sediments into the hypolimnion, while the former is used by the photosynthetic organisms in the epilimnion. The layer in which the methane-oxidizing bacteria are established is quite sharp, and these organisms are the main reason for the sudden depletion of dissolved oxygen at the thermocline. Methane that accumulates in the hypolimnion is rapidly oxidized during the autumn turnover period, when mixing distributes oxygen more uniformly. The methane-oxidizing bacteria however tend to remain stratified. Normally the size of this bacterial population ensures that no methane ever reaches the atmosphere. It does this only in shallow lakes with very little dissolved oxygen.

In paddy fields and marshes, however, a quite different state of affairs exists. Much methane is released to the atmosphere (see Table 1.2).

Nitrogen Metabolism

Eutrophic lakes tend to be rich in fixed nitrogen (organic nitrogen compounds, ammonia, nitrite and nitrate); recent evidence suggests that methane oxidizers in such lakes are responsible for nitrification (oxidation of

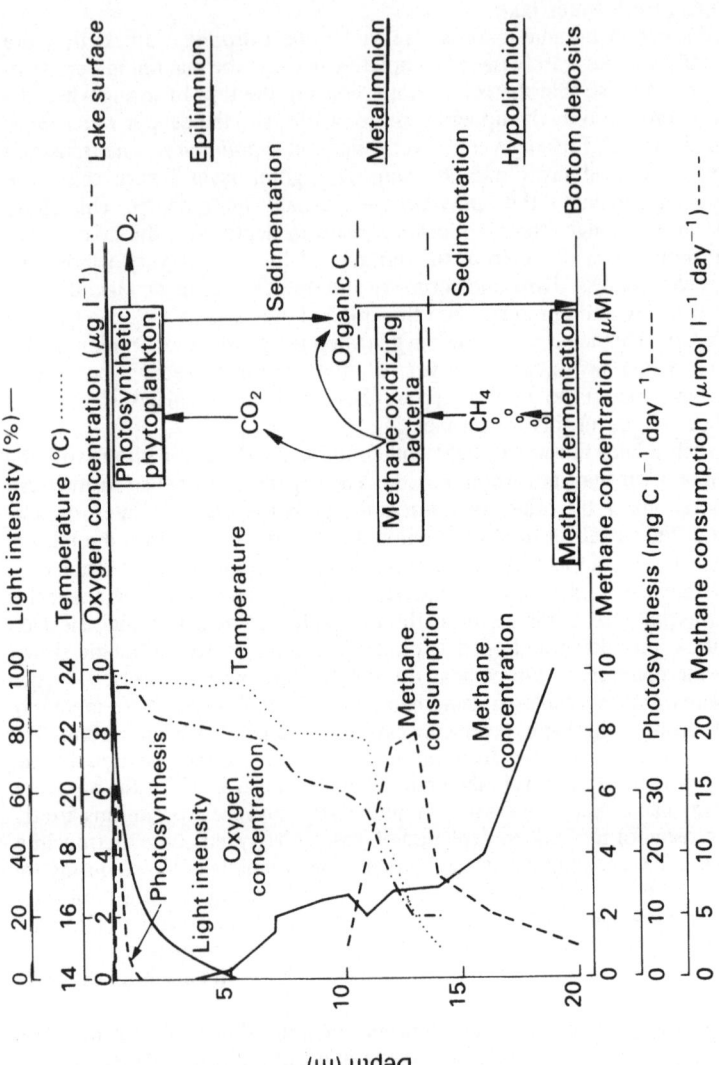

Fig. 1.2 Schematic representation of the circulation of carbon in a stratified eutrophic lake. (The methane data are based on those of Patt et al. (1974).)

ammonia to nitrite and nitrate). In contrast, environments such as paddy fields are extremely deficient in fixed nitrogen. Here the ecosystem is altered by the presence of cyanobacteria (blue-green algae) that can fix both nitrogen and carbon dioxide and evolve oxygen. In this way significant enrichment of the soil with fixed nitrogen occurs which is essential for the growth of rice crops, since nitrogenous fertilizers are often not available. Most methane-oxidizing bacteria (methanotrophs) can fix nitrogen, and they also ought to flourish in the paddy-field ecosystem. The quantitative contribution to nitrogen fixation by methanotrophs in the various different environments has not been assessed, but it is thought to be significant. The pathway by which reduced nitrogen enters cell material also varies with the type of organism. Some use glutamate dehydrogenase, others the glutamine synthetase/glutamate synthase system. Under appropriate conditions the use of one of these pathways instead of another can significantly influence the growth of methylotrophic bacteria (see Chapter 5).

Anaerobic Methane Oxidation

In the last ten years evidence has emerged that in anaerobic marine environments bacteria may be able to live by the anaerobic oxidation of methane using sulphate as electron acceptor:

$$2CH_4 + SO_4^{2-} + 2H^+ \longrightarrow 4H_2 + 2CO_2 + H_2S \tag{1.1}$$

The free energy change for this reaction when the concentrations of sulphate, bicarbonate, sulphide and methane are respectively 10 mM, 30 mM, 0.5 mM and 0.1 mM is $-25\,kJ\,mol^{-1}$, which should be adequate for the organisms to obtain energy for growth. No anaerobic methane depletion has ever been detected in the absence of sulphate. The bacteria involved grow deep down in the sediment, providing that sulphate is available by downward diffusion from the seawater. This anaerobic removal of methane probably means that the figures in Table 1.2 are a serious underestimate of the quantity of naturally produced methane, since they are based on methane reaching the atmosphere. Anaerobic methane oxidation takes place in freshwater habitats to only a small extent because of the low concentrations of sulphate but in the marine environment, where sulphate concentration is about 10 mM, this process is very important.

The organisms involved have not yet been positively identified. Sulphate reducers may be directly involved, although the evidence that they can oxidize methane is not very substantial. Methanogenic bacteria have been shown by Zehnder and Brock (1979) to oxidize methane anaerobically, and it is possible that a consortium of methanogens and sulphate reducers might exist which at high methane concentrations could reverse methanogenesis by inter-species hydrogen transfer (see Chapter 2 for a definition of this term).

$$CH_4 + 2H_2O \longrightarrow CO_2 + 4H_2 \tag{1.2}$$

$$4H_2 + SO_4^{2-} \longrightarrow S^{2-} + 4H_2O \tag{1.3}$$

This would be the opposite state of affairs to that which occurs in the rumen (Chapter 2), where methanogens remove rather than produce hydrogen.

Summary

Many reduced C_1 compounds occur in nature (the most important are listed in Table 1.1). Some species of micro-organisms can grow on these compounds as sole carbon and energy source. Such organisms are called methylotrophs. They occupy an intermediate biological position between autotrophs and heterotrophs. Some methylotrophs can *only* grow on C_1 compounds and are said to be obligate methylotrophs. Those able to grow on methane are called methanotrophs. Methanotrophs and other methylotrophs are in large measure responsible for the oxidation of methane produced by the methanogenic bacteria (Chapter 2), and so convert reduced C_1 compounds into CO_2, thus making it available to photosynthetic organisms. This specialized branch of the carbon cycle has been named the methane cycle (Fig. 1.1). Methanogenic bacteria are those bacteria that live in specialized anaerobic habitats and convert large quantities of organic carbon into methane.

References

ANDERSON, J.W. (1980). *Bioenergetics of Autotrophs and Heterotrophs.* Edward Arnold, London.

EHHALT, D.H. (1976). The atmospheric cycle of methane. In *Symposium on Microbial Production and Utilization of Gases* (H_2, CH_4, CO), pp. 13–22. Edited by H.G. Schlegel, G. Gottschalk and N. Pfennig. Akademie der Wissenschaften, Göttingen.

HANSON, R.S. (1980). Ecology and diversity of methylotrophic organisms. *Advances in Applied Microbiology* **26**, 3–39.

PATT, T.E., COLE, G.C., BLAND, J. and HANSON, R.S. (1974). Isolation and characterization of bacteria that grow on methane and organic compounds as sole sources of carbon and energy. *Journal of Bacteriology* **120**, 955–964.

WOLFE, R.S. and HIGGINS, I.J. (1979). Microbial biochemistry of methane—A study in contrasts. In *Microbial Biochemistry* (International Review of Biochemistry, vol. 21), pp 267–353. Edited by J.R. Ouayle. University Park Press, Baltimore.

ZEHNDER, A.J.B. and BROCK, T.D. (1979) Methane formation and methane oxidation by methanogenic bacteria. *Journal of Bacteriology* **137**, 420–432.

2 Physiology and Biochemistry of Methane-Producing (Methanogenic) Bacteria

The Role of Methanogenic Bacteria in Nature

Only some species of bacteria and a few protozoa can live permanently in the absence of oxygen. If a habitat is rich in biodegradable organic compounds, a nitrogen source and other essential elements, but lacks oxygen, (for example the mud at the bottom of a lake) then an anaerobic microbial population will develop that may contain methanogenic bacteria. Not all such environments are suitable for methanogens. If high concentrations of sulphate or oxidized nitrogen compounds are present, different anaerobes will predominate which under these conditions have a selective advantage over methanogens.

In methanogenic habitats of the type mentioned in Chapter 1, the methanogens are the terminal organisms in a microbial food chain that depends on three other groups of bacteria for the total breakdown of biological polymers into methane, carbon dioxide and (in some habitats) acetate. Such polymers include all the components of living tissue—proteins, lipids and carbohydrates, including cellulose, starch, pentosan and pectin. Lignin is one important compound that is not biodegradable in anaerobic conditions. Some of these anaerobic habitats are highly selective. In some cases the selective pressure is applied by man, who in certain circumstances may wish for virtual quantitative conversion of organic material into gaseous end products. An example of this type of selective environment is the anaerobic sewage sludge digester, widely used in sewage disposal (see Chapter 5) where the end products are methane and CO_2. A naturally occurring selective habitat is the rumen (see Chapter 1), where part of the mixed microbial population breaks down cellulose and other plant biopolymers under strictly anaerobic conditions to give fatty acids, CO_2 and hydrogen gas (dihydrogen, H_2). The fatty acids are used by the host animal, and the CO_2 and hydrogen by the methanogenic bacteria. The methanogens play an important role in directing the rumen fermentation in the direction of fatty acid and acetate formation (see below).

The breakdown of biopolymers in an anaerobic environment requires four types of bacteria which in most anaerobic ecosystems co-exist as a complex mixed population. Only a limited amount of information can be obtained about such a community from studies using pure bacterial cultures. The first stage of anaerobic biodegradation (Fig. 2.1) mainly involves putrefactive Clostridia in the anaerobic sludge digester, whereas bacteria of the genera *Bacteroides*, *Selenomonas* and *Butyrovibrio* carry out this process in the rumen. These bacteria, using extracellular enzymes, hydrolyse the biopolymers protein, carbohydrate and lipid into respectively amino acids, sugars and fatty acids. The carbohydrates starch and cellulose are quantitatively the most important of these polymers. These are then broken down further in

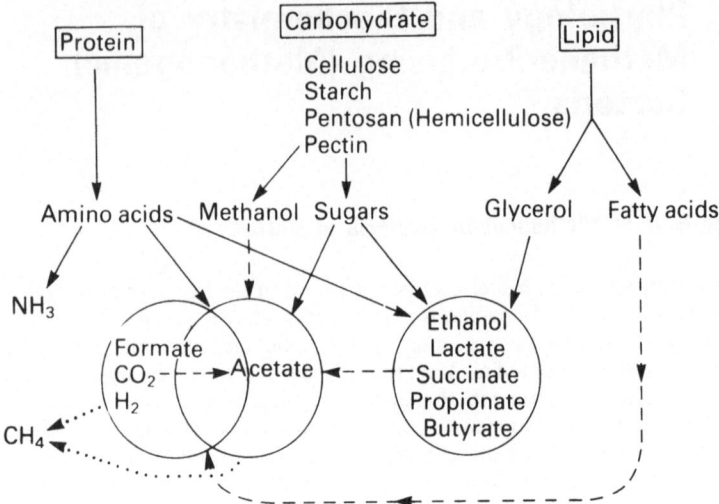

Fig. 2.1 Anaerobic breakdown of biopolymers to give rise to methane. $\longrightarrow$ Primary fermentation processes, $--\rightarrow$ secondary fermentation processes, $\cdots\rightarrow$ tertiary (methanogenic) fermentation phase.

this *primary* fermentative stage to yield acetate, formate, CO_2 and H_2, and in addition ethanol, lactate, succinate, propionate and butyrate. The latter five compounds are typical fermentation products of bacteria in which highly reduced end products have to be formed to act as a 'sink' for excess electrons derived from the growth substrate, i.e. the reduction level of the end products has to match that of the growth substrate. An example might be an organism fermenting glucose to acetate, formate and ethanol (Equation 2.1).

$$C_6H_{12}O_6 + 3ADP + 3P_i \longrightarrow CH_3CH_2OH + CH_3COOH \\ + 2HCOOH + 2H_2O + 3ATP \quad (2.1)$$

If this organism could reoxidize its reduced nicotinamide nucleotides using protons as electron acceptors, with the formation of dihydrogen, then it would have no need to form ethanol. This would be replaced in the fermentation products by acetate (Equation 2.2)

$$C_6H_{12}O_6 + 4ADP + 4P_i \longrightarrow 2CH_3COOH + 2H_2 + 2HCOOH \\ + 2H_2O + 4ATP \quad (2.2)$$

and the cell, instead of losing a high energy acetyl-coenzyme A by reduction to ethanol could convert it via acetyl phosphate, yielding an extra ATP, thus increasing its energy yield per mole of glucose by 33%. Under conditions of pure culture, this would not be possible, because the equilibrium constant (K_{eq}) for the reaction:

$$NAD(P)H + H^+ \rightleftharpoons NAD(P)^+ + H_2 \quad (2.3)$$

is 4.2×10^{-4}; that is, massively in the direction of $NAD(P)H + H^+$ rather than of H_2 formation.

In the mixed populations of the rumen and anaerobic sediments there are many bacteria that can utilize dihydrogen. One group of these is the methanogenic bacteria. The others are bacteria, such as *Acetobacterium* and *Clostridium aceticum* (*homoacetogenic* bacteria), which can convert CO_2 and H_2 into acetate. These organisms, by removing dihydrogen as quickly as it is formed pull Equation 2.3 to the right and allow $NAD(P)H$ to be oxidized to $NAD(P)^+$ and H_2. By their presence, not only do these two bacterial groups allow more favourable primary fermentation, but they also allow the development of other bacteria including a group known as *obligate proton reducers*. These various groups of bacteria can ferment ethanol, lactate succinate, propionate and butyrate to acetate and H_2 (the *secondary* fermentative stage). Since these bacteria can use no other electron acceptor except protons, they can only develop when the dihydrogen in the medium is removed by methanogenic or homoacetogenic bacteria. An example of such an organism is the S-organism that grows in symbiotic association with *Methanobacterium bryantii*, an association which for many years was thought to be a single organism, '*Methanobacillus omelianskii*'.

The *tertiary* and final stage in this food chain involves the methanogenic bacteria, which were mentioned in the previous paragraph. They obtain energy by the oxidation of dihydrogen under anaerobic conditions using CO_2 as electron acceptor (Equation 2.4).

$$HCO_3^- + H^+ + 4H_2 \longrightarrow CH_4 + 3H_2O \quad \Delta G^{0'} = -135.6 \, \text{kJ mol}^{-1} \quad (2.4)$$

Living cells are capable of using the standard free energy change of this reaction to obtain ATP, although less free energy is available (per eight electrons transferred) than from the reduction of sulphate to sulphide ($\Delta G^{0'} - 151.9$ kJ per 8 electrons, nitrate to nitrite (-652.8 kJ per 8 electrons) or oxygen to water (-949.0 kJ per 8 electrons). Our limited knowledge of the physiology of these organisms is discussed in this chapter. By their existence in anaerobic habitats, they make conditions more favourable for the primary and secondary (acetogenic) fermentations by removing dihydrogen. This important phenomenon is called *interspecies hydrogen transfer*.

The gaseous end products of biodegradation in the rumen are CO_2 (60 to 70% of the gas) and methane (30 to 40% of the gas). This gas is belched up by the ruminant at a rate of 60 to 80 l day^{-1}. The other products formed are acetate (47 to 60% of the acid), propionate (18 to 23%) and butyrate (19 to 29%), which are absorbed into the ruminant's blood stream and used as its principal energy source. The protein needed by the ruminant is mainly obtained by digesting those bacterial cells that overflow from the rumen and pass further along the animal's alimentary canal. In this way amino acids, vitamins and all other necessary nutrients are obtained by the ruminant, whose bodily metabolism is based on the oxidation of fatty acids rather than carboydrates.

The habitats in which methanogenic bacteria are found were mentioned in Chapter 1. With the exception of the thermophilic strains, methanogens of all genera can be isolated from both type A (sediments) and type B (rumen) habitats.

13

Classification and Characteristics of Methanogenic Bacteria

The traditional morphological classification of the methanogens has recently been discarded in favour of a taxonomy based on the structure of the 16S ribosomal RNA. This molecule has changed very slowly during evolution, so any significant differences that are observed must indicate a very long evolutionary history. Study of the structure of 16S rRNA in the methanogens has revealed that there is a very great diversity among the methanogens themselves; they are not a homogeneous and closely related group of bacteria (Table 2.1).

The differences between the Methanobacteriales on the one hand and the Methanococcales on the other (see Table 2.1) are as great as the differences between Gram-positive and Gram-negative Eubacteria, which means that the divergences are probably at least as old. It appears that methanogens are completely distinct from virtually all other known bacteria with the exception of the extreme halophiles (genus *Halobacterium*) and the thermoacidophiles (genera *Thermoplasma* and *Sulfolobus*). To accommodate this discovery, it has been proposed (Stackebrandt & Woese, 1981) that a separate primary kingdom or *urkingdom* be recognized among the prokaryotes to include the methanogens, the extreme halophiles and the thermoacidophiles. This

Table 2.1 Classification of Seven Genera of Methanogenic Bacteria (Based on Comparative 16S rRNA Structure)

		Morphology	Gram reaction	Cell wall	Growth substrates
Order I	*Methanobacteriales*				
Family I	*Methanobacteriaceae*				
Genus 1	*Methanobacterium*	Long rods to filaments	+	Pseudomurein	H_2/CO_2 Formate
Genus 2	*Methanobrevibacter*	Short rods to lancet cocci	+	Pseudomurein	H_2/CO_2 Formate
Order II	*Methanococcales*				
Family I	*Methanococcaceae*				
Genus 3	*Methanococcus*	Irregular and small	−	Protein subunits with trace of glucosamine	H_2/CO_2 Formate
Order III	*Methanomicrobiales*				
Family I	*Methanomicrobiaceae*				
Genus 4	*Methanomicrobium*	Short rods	−	Protein subunits	H_2/CO_2 Formate
Genus 5	*Methanogenium*	Irregular small cocci	−	Protein subunits	H_2/CO_2 Formate
Genus 6	*Methanospirillum*	Short to long wavy spirilla	−	Protein subunits Protein sheath	H_2/CO_2 Formate
Family II	*Methanosarcinaceae*				
Genus 7	*Methanosarcina*	Pseudosarcina	+	Heteropoly-saccharide	H_2/CO_2 Methanol Acetate Methylamine

urkingdom would be called the *archaebacteria*. All the remaining bacteria, cyanobacteria and mycoplasmas would be in the urkingdom *eubacteria*. Eukaryotes have a mixed ancestry: their intracellular organelles (chloroplasts and mitochondria) are thought to have arisen from eubacteria, and their cytoplasmic component with its characteristic 18S ribosomal RNA from a hypothetical parental urkingdom, the *urkaryotes*. How far these views will commend general acceptance remains to be seen, but the methanogens do have many unique biological properties (Table 2.2). The revised genera of methanogens are listed in Table 2.1. The pseudomurein of the Methanobacteriales contains only L-amino acids (D-amino acids are totally absent) and N-acetyltalosaminuric acid in place of muramic acid. The polar lipids are nonsaponifiable ether-linked polyisoprene lipids (i.e. they contain long-chain primary alcohols rather than fatty acids), whereas the neutral lipids are mainly squalene derivatives, quite different from eubacterial lipids (see Glossary for further information).

Table 2.2   Differences between Eubacteria and Methanogenic Bacteria

Eubacteria	Methanogens
1. Cell wall contains murein	1. Murein absent. Pseudomurein found in Methanobacteriales
2. Coenzyme M absent	2. Coenzyme M present in all species
3. Electron carriers are flavoproteins, quinones, cytochromes and ferredoxin	3. Contain unusual electron carriers — coenzyme F_{420}, an unusual flavin dervative — factor F_{430}, a nickel tetrapyrrole derivative Quinones and ferredoxin probably absent. Cytochromes and flavins rare
4. Membrane lipids are phosphatidyl derivatives and mono-, di- and tri-acylglycerols, all containing fatty acids.	4. Membrane lipids are C_{20} phytanyl and C_{40} biphytanyl glycerol *ethers* and isoprene hydrocarbons (mainly squalene)
5. No atractyloside-sensitive adenine nucleotide translocase has been found	5. Membranes contain an atractyloside-sensitive adenine nucleotide translocase
6. CO_2-fixation involves the Calvin cycle (except for the genus *Chlorobium*)	6. CO_2-fixation does not involve the Calvin cycle, nor the reductive tricarboxylic acid cylce of *Chlorobium*
7. RNA polymerase consists of a β' $\beta\alpha_2$ δ set of subunits	7. RNA polymerases do not have this structure
8. Transfer RNA contains ribothymine in the TψC loop	8. TψC sequence absent from tRNA
9. Peptide elongation factor EF-G not sensitive to diphtheria toxin	9. Peptide elongation factor is ADP-ribosylated by diphtheria toxin

Growth of Methanogens

The isolation, handling and growth of methanogenic bacteria are extremely difficult unless the appropriate rigorous precautions are taken to ensure that no oxygen whatsoever is present. This includes the use of anaerobic roll tubes for solid media and pressurized culture vessels for liquid and large-scale cultivation. These special techniques are described in the review of Balch *et al.* (1979).

The best substrate for the growth of virtually all methanogens is dihydrogen + CO_2 (H_2/CO_2). The energy available from the coupled reduction of CO_2 by H_2 (Equation 2.4) is sufficient to support reasonable growth. Formate can also be metabolized by most species of methanogens. It is probably first oxidized by a formate dehydrogenase to $H_2 + CO_2$. The metabolically most versatile species are strains of *Methanosarcina*, which can grow on the above compounds and also on acetate, methanol, methylamine, dimethylamine, trimethylamine and ethyldimethylamine. In addition, some methanogens can oxidize carbon monoxide (CO) and convert it to methane, and a few strains can grow on it. Growth on CO_2 as carbon source is *autotrophy*, but the autotrophic growth of the methanogens is totally different from that of virtually all phototrophs and chemoautotrophs because it does not involve the ribulose bisphosphate (Calvin) cycle (Chapter 3). Some species of methanogen require vitamins in addition to H_2/CO_2, and some also require acetate or methylbutyrate. *Methanobacterium thermoautotrophicum* is a thermophilic methanogen (the optimum growth temperature is 65–70°C). As it has no vitamin or other additional requirements and grows relatively rapidly (doubling time 3–5 h at 65°C) it has been the subject of much recent research.

Biochemistry of Methane Formation

The dihydrogen-dependent reduction of CO_2 to methane requires an anaerobic electron transport pathway. The exact nature of the electron carriers involved in this pathway, and the extent to which membrane-bound and soluble components interact have not yet been determined. The initial reduction of CO_2 to the level of formaldehyde is thermodynamically unfavourable, whereas further reduction of formaldehyde to yield methane is very favourable, and the standard free energy change of Equation 2.4 would theoretically support the formation of two molecules of ATP. The electron carriers present in methanogenic bacteria are unusual. Cytochrome *b* has been reported in *Methanosarcina barkeri* and membrane-bound flavin-adenine dinucleotide in *Methanobacterium bryantii*, but otherwise the electron carriers so far detected seem to be unique to methanogenic bacteria. Most of these components are soluble, but it seems likely that in the cell they function in a membrane-bound form. Many of them are inactivated by oxygen, and our recent progress in characterizing them has been helped by purification under anaerobic conditions.

The initial step in the reduction process involves the only membrane-bound component so far characterized, hydrogenase, which converts H_2 gas into electrons and protons (see Equation 2.6). The redox potential for this reaction is $-414\,mV$. This hydrogenase contains nickel, but not in the form of factor F_{430}. Although the redox potentials of some of the other purified components have been measured, in many cases their function is uncertain, and in consequence it is not yet possible to propose an order for the electron carriers in the overall reduction process. The cofactors so far described are coenzyme $M(E'_0 - 193\,mV)$, component B, coenzyme F_{420} ($E'_0 - 373\,mV$), factors F_{342} (blue-fluorescent) and F_{430} (yellow, non-fluorescent), YFC (yellow-fluorescent), and B'_0 ($E'_0 - 450\,mV$), CDR (carbon dioxide reduction) factor and compound C_1-X-T (Keltjens & Vogels, 1981). The structure and biochemical role of two of these, coenzyme M and coenzyme F_{420} have been elucidated, but our knowledge of the overall pathways and the role of the other cofactors in metabolism is still rather tenuous. The structure of B_0 has recently been shown to be that of a pteridine derivative (named *methanopterin*) and YFC is its carboxy-7,8-dihydro derivative. Factor F_{430} is an unusual tetrapyrrole derivative containing nickel.

Role of Coenzyme F_{420}. Although not established for certain, it seems probable that the electron carrier coenzyme F_{420} does not play a direct role in methanogenesis. It is a deazaflavin with a long side chain ending in two glutamate residues (Fig. 2.2). It accepts two electrons and has E'_0 of $-373\,mV$. Its main role is as an electron acceptor for hydrogenase, formate dehydrogenase, pyruvate and 2-oxoglutarate dehydrogenases and NADPH, so that it can function as part of an NADPH-dependent hydrogenase system, as part of a formate hydrogenlyase system ($HCOOH \rightleftharpoons H_2 + CO_2$) or as part of a pyruvate or 2-oxoglutarate synthase system (Fig. 2.3). It seems in fact to play a role analogous to that played by ferredoxin in other anaerobes. Ferredoxin has not been found in pure cultures of methanogens.

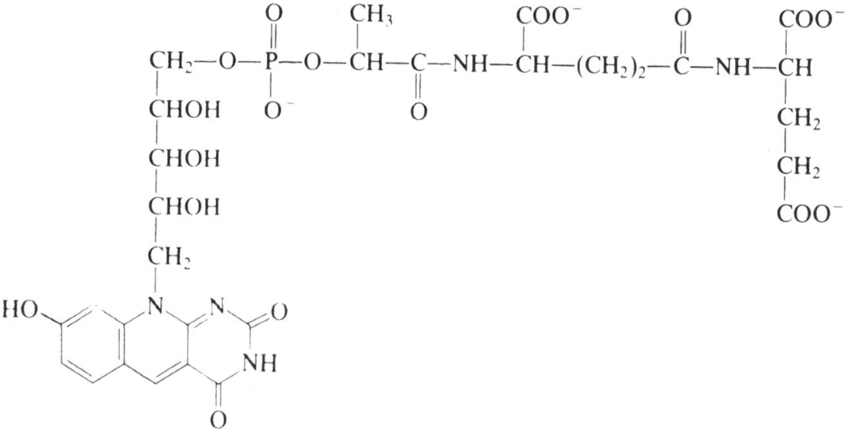

Fig. 2.2 Structure of coenzyme F_{420}, the N-(N-L-lactyl-γ-L-glutamyl)-L-glutamic acid phosphodiester of 7.8-didemethyl-8-hydroxy-5-deazariboflavin 5'-phosphate.

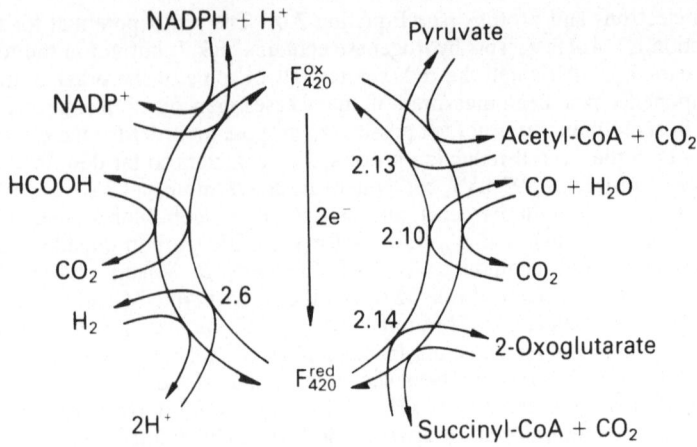

Fig. 2.3 Some of the redox reactions involving coenzyme F_{420} in methanogenic bacteria. The numbers refer to equations in the text.

Coenzyme F_{420} is not unique to methanogenic bacteria. It has recently been found in *Streptomyces griseus*.

Carbon Dioxide Reduction. Very early experimental work showed that methanol, formaldehyde and formate were not intermediates in methane formation, so that the reduction process must occur with the C_1 compounds bound to a carrier. One carrier was discovered in 1974 and given the name coenzyme M. It was shown to be 2-mercaptoethanesulphonic acid, $HSCH_2CH_2SO_3^-$, and the functional part of the molecule is the -SH group. Coenzyme M occurs in all methanogenic bacteria, and appears to be unique to this group of organisms. One organism, *Methanobrevibacter ruminantium*, is unable to synthesize coenzyme M and requires it as a vitamin for growth. An early labelled reduction product of $^{14}CO_2$ that can be detected in whole cells or extracts of methanogenic bacteria is methyl-coenzyme M, $CH_3SCH_2CH_2SO_3^-$ (abbreviated CH_3–S–CoM). This compound is the substrate for the final step in methane production catalysed by *methyl-coenzyme M reductase*.

When CH_3–S–CoM is added to extract of a methanogen under an atmosphere of H_2/CO_2, there is a great stimulation of methane formation (12 mol of methane formed per mol of CH_3–S–CoM added). This demonstrates that the methyl-coenzyme M reductase reaction is coupled to the activation and reduction of CO_2. Thus the final step seems to generate an intermediate that is involved in the primary step of CO_2 activation, so that we have a cyclic system as shown in Fig. 2.4. Hydroxymethyl-coenzyme M is reduced by crude extracts and may also be an intermediate in methane formation. Whether coenzyme M functions earlier in the cycle or whether a different factor, indicated in Fig. 2.4 by X, is the carrier, is not as yet known.

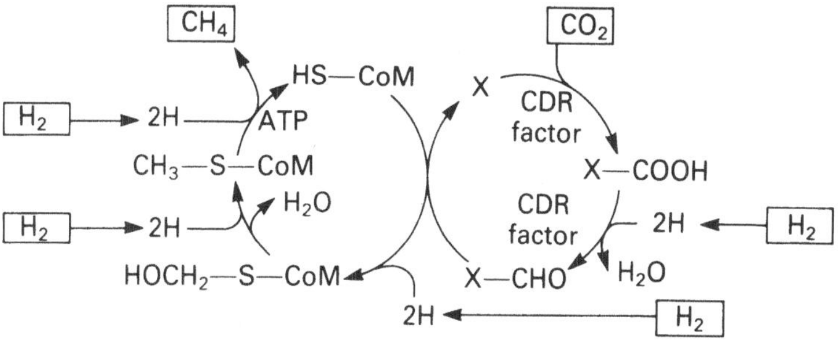

Overall reaction: $CO_2 + 4H_2 \longrightarrow CH_4 + 2H_2O$

Fig. 2.4 Formation of methane by reduction of CO_2. Reactants and final products are shown in boxes. X is an unknown carrier. (Modified from J.A. Romesser, Ph.D. thesis (1978) University of Illinois, Urbana, IU., USA, by permission.)

Methyl-Coenzyme M Reductase. The final step in methane reduction (Equation 2.5) has been extensively studied. Three components A, B and C are involved in addition to ATP, Mg^{2+} ions and the substrates

$$CH_3—S—CoM + H_2 \xrightarrow[A.B.C]{ATP, Mg^{2+}} CH_4 + HS—CoM \qquad (2.5)$$

Of these components, A is a membrane-bound, oxygen-sensitive hydrogenase (Equation 2.6)

$$H_2 \rightleftharpoons 2H^+ + 2e^- \qquad (2.6)$$

B is an oxygen-labile low molecular weight coenzyme, and C is the methylreductase itself. The hydrogenase can be replaced by purified NADPH-coenzyme F_{420} oxidoreductase, NADPH and coenzyme F_{420} (Equation 2.7)

$$CH_3—S—CoM + NADPH + H^+ \xrightarrow[\substack{B.C.F_{420} \\ \text{oxidoreductase}}]{ATP, Mg^{2+}} CH_4 + HS—CoM + NADP^+ \qquad (2.7)$$

The methylreductase has recently been purified. It has a molecular weight of 300 000 and consists of three different kinds of subunit. The protein is yellow, but the nature of the chromophore is not yet known. It may be factor F_{430}. The enzyme system has been found in all methanogens. It is inhibited by tripolyphosphate (which binds the Mg^{2+} ions).

Recent work with *Methanosarcina barkeri* has shown that methyl-coenzyme M can be formed in cell-free extracts of that organism from methanol (Equation 2.8) or methylamine

$$CH_3OH + HS—CoM \xrightarrow{ATP, Mg^{2+}} CH_3—S—CoM + H_2O \qquad (2.8)$$

The reaction requires ATP. This ATP requirement, like that for Equation 2.5, is catalytic, i.e. 1 mol of ATP will promote the formation of 5–15 mol methane. Thus the role of ATP may be to phosphorylate or adenylylate one of the enzymes of the system rather than to act as a substrate.

19

It has recently been shown also that in *Methanobacterium thermoautotrophicum* growing on H_2/CO_2, the hydrogen atoms in the methane formed arise exclusively from water and not from dihydrogen. Thus during methanogenesis, H_2 gas is serving only as a source of electrons. Several methanogenic bacteria, particularly *Methanosarcina barkeri*, contain such large quantities of cobalamin (vitamin B_{12}) derivatives, that the cells appear red. Moreover chemically synthesized methylcobalamin can be converted into methane by crude cell extracts. However, methylcobalamin is a very good non-enzymic methylating agent, and present views are that cobalamin compounds are not involved in methane formation during growth on H_2/CO_2, although they may be involved during methanogenesis from methanol.

Carbon Monoxide Metabolism By Methanogens. Most methanogens can metabolize carbon monoxide (CO) to methane during growth on H_2/CO_2. This has been shown for *Methanobacterium formicicum, M. thermoautotrophicum, Methanobrevibacter ruminantium, M. arboriphilus* and *Methanosarcina barkeri*. In addition, *Methanobacterium thermoautotrophicum* can grow slowly on CO (30% CO/70% He) in place of H_2/CO_2. The stoicheiometry for this is given in Equation 2.9

$$4CO + 2H_2O \longrightarrow CH_4 + 3CO_2 \tag{2.9}$$

This process proceeds via the oxidation of CO to CO_2 which is then reduced to methane. This oxidation must involve a dehydrogenase, and a CO dehydrogenase has been identified in *Methanobacterium thermoautotrophicum*. The enzyme uses coenzyme F_{420} as electron acceptor (Equation 2.10).

$$CO + H_2O + F_{420}^{ox} \longrightarrow CO_2 + F_{420}^{red} \tag{2.10}$$

Ferredoxin, NAD^+ or $NADP^+$ would not replace coenzyme F_{420}. This enzyme has not yet been purified, so it is not yet established whether it contains nickel as do the CO dehydrogenases from *Clostridium pasteurianum* and *C. thermoaceticum*. The electron acceptor of the last two enzymes is not coenzyme F_{420} but ferredoxin.

Metabolism of *Methanosarcina barkeri*. *Methanosarcina barkeri* is interesting because of its ability to use acetate, methanol and methylated amines as sole carbon and energy source instead of H_2/CO_2. This makes it (in the case of methanol and methylated amines) an anaerobic methylotroph. It has been shown that in the fermentation of methanol and methylated amines, the methyl group is transferred intact into methane: there is thus no evidence of the C_1 substrate being oxidized first to CO_2. As we have seen, this reduction probably proceeds via Equations 2.8 and 2.5. Methane is also formed mainly from the methyl group of acetate. No external electron donors are required and the stoicheiometry is (Equation 2.11):

$$CH_3COO^- + H_2O \longrightarrow CH_4 + HCO_3^- \tag{2.11}$$

Nonetheless, the methyl group of acetate can also be oxidized under these conditions.

Energy Coupling and ATP Formation

Methanogenesis *in vivo* is probably a function of the intracellular membranes. *In vitro* work on methanogenesis as described above involved the use of crude extracts which contain large amounts of broken membranes. Evidence has been obtained that methanogenesis can take place in washed membrane vesicles without addition of ATP, coenzyme M or soluble enzymes. It has also been shown that hydrogenase and ATP synthetase are membrane-bound. Formation of ATP in membrane vesicles could be driven either by hydrogen oxidation or by a potassium gradient. These processes were inhibited by uncoupling agents, by nigericin or by the ATPase inhibitor *N,N'*-dicyclohexylcarbodiimide. Methyl-coenzyme M will stimulate methanogenesis in membrane preparations without apparently being reduced to methane. This methanogenesis is also sensitive to uncoupling agents. ATP synthesis in membranes of *Methanobacterium thermoautotrophicum* is sensitive to three different inhibitors of the mitochondrial adenine nucleotide translocase (the exchange mechanism by which ATP passes through the inner mitochondrial membrane in eukaryotes). The presence of such a translocase in *Methanobacterium thermoautotrophicum* has recently been demonstrated. Translocases of this type, sensitive to the inhibitor atractyloside have not previously been found in prokaryotes (see Table 2.2).

Synthesis of Cell Constituents in Methanogenic Bacteria

The problem of studying autotrophic CO_2-fixation in methanogenic bacteria is the relatively small amount of CO_2 which is 'fixed' into cell material in comparison with the amount that is reduced to methane. Careful examination of both *Methanosarcina barkeri* and *Methanobacterium thermoautotrophicum* however has revealed the total absence of key enzymes of the Calvin cycle, the hexulose phosphate cycle and serine pathway (see Chapter 3 for details of these pathways). Moreover, key enzymes of the reductive tricarboxylic acid cycle which is thought to operate for autotrophic CO_2-fixation in the green photosynthetic bacteria of the genus *Chlorobium* are also lacking in the methanogens (see below). An examination of $^{14}CO_2$-fixation products in *Methanobacterium thermoautotrophicum* revealed that alanine and aspartate appeared as the earliest labelled products, with significant amounts of radioactivity also appearing in glutamate and coenzyme M derivatives. Two other significant unidentified radioactive products were C_1–X–T and YFC. The latter has recently been identified as a carboxylated pteridine derivative (see above). [^{14}C]Acetate in *Methanobacterium thermoautotrophicum* gave rise to labelled alanine and succinate, but no label appeared in coenzyme M or pteridine derivatives. In contrast, with *Methanosarcina barkeri*, the first assimilation products of [^{14}C]methanol were coenzyme M derivatives and C_1–X–T.

Study of cell-free extracts of *Methanobacterium thermoautotrophicum* has demonstrated three enzymes capable of fixing CO_2: phosphoenolpyruvate carboxylase (EC 4.1.1.31) (see Equation 2.12):

$$\text{Phosphoenolpyruvate} + CO_2 \longrightarrow \text{Oxaloacetate} + P_i \qquad (2.12)$$

pyruvate synthase and 2-oxoglutarate synthase, the two latter reactions requiring reduced coenzyme F_{420} as electron donor rather than ferredoxin (Equations 2.13, 2.14)

$$CH_3CO\!-\!SCoA + CO_2 + F_{420}^{red} \longrightarrow CH_3COCOOH + CoASH + F_{420}^{ox} \quad (2.13)$$

$$Succinyl\!-\!SCoA + CO_2 + F_{420}^{red} \longrightarrow 2\text{-Oxoglutarate} + CoASH + F_{420}^{ox} \quad (2.14)$$

The evidence thus points to the carboxylation of acetyl-coenzyme A as being a key fixation reaction for CO_2 (see Fig. 2.5). How the acetyl-coenzyme A acceptor is regenerated in this scheme has not yet been established. The absence of isocitrate dehydrogenase in *Methanobacterium thermoautotrophicum* makes unlikely the occurrence of a CO_2-fixation pathway of the kind thought to operate in the green photosynthetic bacteria of the genus *Chlorobium* (Trüper, 1981), in which CO_2 fixed into 2-oxoglutarate is reduced via isocitrate and a reversed (reductive) tricarboxylic acid cycle. Possibly acetyl-coenzyme A is synthesized *de novo* from CO_2 via reactions of

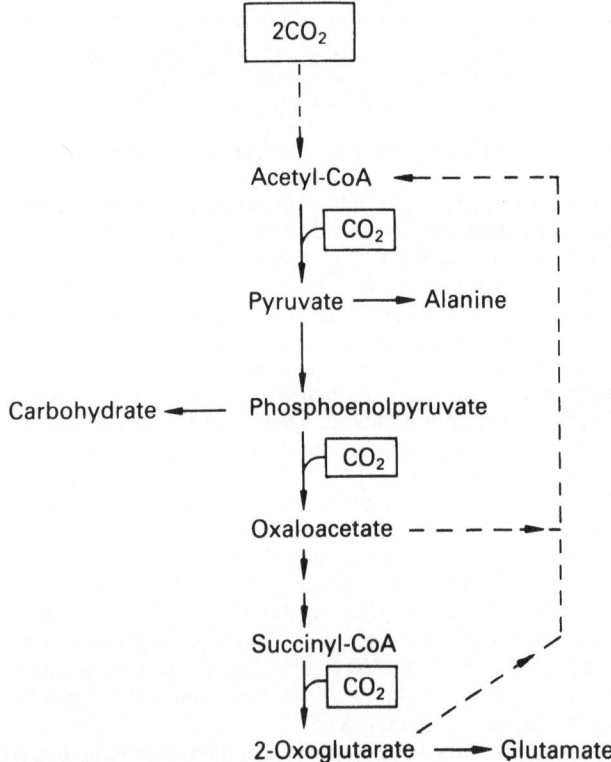

Fig. 2.5 Fixation of CO_2 in *Methanobacterium thermoautotrophicum*. The broken line signifies possible reactions for the formation of acetyl-coenzyme A that have not yet been demonstrated.

a type already known in Clostridia (although free acetate is probably not involved). In *Methanosarcina barkeri* acetate metabolism is different because significant levels of isocitrate dehydrogenase are present and the carboxyl group of $[1-^{14}C]$acetate will label the C-5 of glutamate. Differences of this magnitude are not unexpected in what is a relatively diverse prokaryotic urkingdom.

Conclusions

This relatively long chapter demonstrates how rapidly our knowledge of the biochemistry and physiology of the methanogenic bacteria has developed in the last ten years. Outstanding points yet to be solved include the nature of the reduction reactions between CO_2 and methyl-coenzyme M, the nature of the membrane-associated ATP-synthesis system and the details of how acetyl-coenzyme A can be formed from CO_2. Identification of a number of new redox and other cofactors and assessment of their role in these reactions also remains to be achieved. All that has been learned about the biochemistry of this unique bacterial group reinforces the view that they represent, on evolutionary criteria, a very ancient group of prokaryotes, with distinct differences from the eubacteria.

Summary

Most methanogenic bacteria obtain their energy for growth from the anaerobic oxidation of H_2 using CO_2 as the electron acceptor. They play a very important ecological role in the anaerobic biodegradation of organic material. This is because they enable other bacteria to develop which can ferment organic material with the formation of H_2 in various ecosystems (of which the rumen is one example). These bacteria can only develop if the partial pressure of H_2 is kept extremely low by its continuous removal by the methanogens, a process called inter-species hydrogen transfer. The methane so formed plays an important role in the carbon cycle (Chapter 1). Methanogenic bacteria belong to the archaebacteria, a group of prokaryotes that is so different from the remaining prokaryotes that it almost certainly represents a different evolutionary urkingdom (primary kingdom) of living things. These differences are summarized in Table 2.2. Possibly such ancient organisms have survived independently without much change because their habitats resemble those on the primitive earth before the cyanobacteria gave rise to an aerobic atmosphere. The other archaebacteria are the genera *Halobacterium*, *Thermoplasma* and *Sulfolobus*, all organisms with peculiar habitats. The seven identified genera of methanogens are extremely diverse: the differences between some groups are as great as the differences between Gram-positive and Gram-negative bacteria. Methanogens require special techniques for their isolation, handling and large-scale growth. They all utilize H_2/CO_2 as growth substrate, and some species will also use formate. *Methanosarcina barkeri* will also use acetate, methanol and methylated amines for growth.

The biochemistry of the methanogens also presents unusual features. Central to the process of methane evolution is coenzyme M, a methyl group carrier that is the immediate precursor of methane. This coenzyme, 2-mercaptoethanesulphonate, is unique to methanogens. They also contain other unusual cofactors—an electron carrier coenzyme F_{420}, which functions in methanogens instead of ferredoxin; a novel pteridine, methanopterin, which may be a C_1-carrier; and a novel nickel-tetrapyrrole cofactor, F_{430}. Methanogenesis is almost certainly catalysed by the membranes of the cell, and membrane vesicles can give rise both to ATP and to methane by an uncoupler-sensitive process. Cell constituents are made via acetyl-coenzyme A which is carboxylated to pyruvate. The metabolic pathway for the generation of the acetyl-coenzyme A is not yet known. Possibly a direct combination of two CO_2 molecules is involved.

References

BALCH, W.E., FOX, G.E., MAGRUM, L.J., WOESE, C.R. and WOLFE, R.S. (1979). Methanogens: Reevaluation of a unique biological group. *Microbiological Reviews* **43**, 260-296.

STACKEBRANDT, E. and WOESE, C.R. (1981). The evolution of prokaryotes. In *Molecular and Cellular Aspects of Microbial Evolution*, 32nd Symposium of the Society for General Microbiology pp. 1-31. Edited by M.J. Carlisle, J.F. Collins and B.E.B. Moseley. University Press, Cambridge.

WOLFE, R.S. (1979). Microbial biochemistry of methane. Part I. Methanogenesis. In *Microbial Biochemistry* (International Review of Biochemistry, vol. 21), pp. 270-300. Edited by J.R. Quayle. University Park Press, Baltimore.

WOLFE, R.S. (1981). Respiration in methanogenic bacteria. In *Diversity of Bacterial Respiratory Systems*, vol. 1, pp. 161-186. Edited by C.J. Knowles. CRC Press, Boca Raton, Florida.

WOLIN, M.J. (1976). Interactions between H_2-producing and methane-producing species. In *Microbial Production and Utilization of Gases* (H_2, CH_4, CO), pp. 141-150. Edited by H.G. Schlegel, G. Gottschalk and N. Pfennig. Academy of Sciences, Gottingen.

ZEIKUS, J.G. (1977). The biology of methanogenic bacteria. *Bacteriological Reviews* **41**, 514-541.

The following references are all to chapters in *Microbial Growth on C_1 Compounds: Proceedings of Third International Symposium*. Edited by H. Dalton. Heyden & Son, London.

ELLEFSON, W.L. and WOLFE, R.S. (1981). Biochemistry of methylreductase and evolution of methanogens, pp. 171-180.

KELL, D.B., DODDEMA, H.J. MORRIS, J.G. and VOGELS, G.D. (1981). Energy coupling in methanogens, pp 159-170.

KELTJENS, J.T. and VOGELS, G.D. ('981). Novel coenzymes of methanogens, pp. 152-158.

MAH, R.A., SMITH, M.R., FERGUSON, T. and ZINDER, S. (1981). Methanogenesis from H_2-CO_2, methanol, and acetate by *Methanosarcina*, pp. 131-142.

TRUPER, H.G. (1981). Versatility of carbon metabolism in the phototrophic bacteria, pp. 116-121.

3 Physiology and Biochemistry of Methylotrophic Bacteria

In this chapter we shall be considering the pathways by which methylotrophic bacteria can obtain energy and carbon for growth by metabolism of reduced C_1 compounds (such as those in Table 1.1) as their sole carbon source. Yeasts will be discussed in Chapter 4. In contrast to the strictly anaerobic methanogenic bacteria discussed in Chapter 2, the methylotrophs, except for certain photosynthetic bacteria, are aerobes. To some extent the word methylotroph is misleading, for, as we have seen in Chapter 1, some of the reduced C_1 compounds supporting methylotrophic growth do not contain any methyl groups (e.g. formate and carbon monoxide).

Basically we may divide the metabolic transformations of C_1 compounds into two areas as shown in Fig. 3.1. They are

1. the pathways by which carbon enters cell material, which we may call *assimilatory pathways*, and
2. the pathways by which the growth substrates are oxidized to yield energy, which we shall call *dissimilatory pathways*.

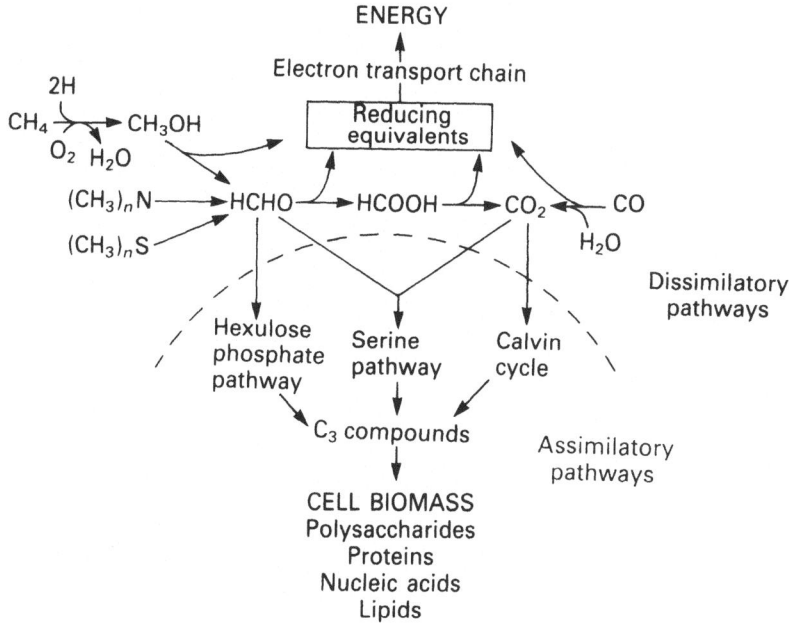

Fig. 3.1 Growth of bacteria on reduced C_1 compounds.

This distinction is particulary useful in the case of methylotrophic metabolism, where there are clear metabolic branch-points between assimilatory and dissimilatory pathways. In heterotrophic growth on multicarbon compounds, the pathways of substrate-breakdown are tightly integrated with the biosynthetic pathways of the cell. As we foresaw in Chapter 1, the metabolic pathways in both the above two areas show several unusual features that make the methylotrophs a unique group from both a physiological and a biochemical point of view.

Of the reduced C_1 compounds given in Table 1.1, only two oxidation products are converted into cell material, namely formaldehyde and CO_2 (Fig. 3.1). We can divide methylotrophs into three classes according to their physiology. In the first class are the *heterotrophic methylotrophs*, which apart from reduced C_1 compounds can also use a wide variety of multicarbon compounds for growth. These organisms are neither photosynthetic nor chemoautotrophic. They assimilate all or some of their carbon at the level of formaldehyde. The second class, the *autotrophic methylotrophs* oxidize their reduced C_1 growth substrate to CO_2, and grow on the latter. This group of methylotrophs (called by Zatman, 1981, *pseudomethylotrophs*) is very diverse. It includes chemoautotrophic bacteria, which in addition to growing methylotrophically can also grow by oxidation of hydrogen or reduced sulphur compounds, as well as organisms such as *Pseudomonas oxalaticus* that grow on formate by oxidizing it to CO_2 and assimilating the latter. Some photosynthetic bacteria also fall into this class, many representatives of which can also grow heterotrophically. The term 'autotrophic' is used here to indicate the ability to use CO_2 as sole carbon source. It does not imply obligate autotrophy. Both autotrophic and heterotrophic methylotrophs have purely autotrophic or purely heterotrophic modes of existence open to them as an alternative to methylotrophy: They are thus *facultative* methylotrophs. The third class of methylotrophs, the *obligate methylotrophs*, do not have alternative modes of existence. They can only grow on reduced C_1 compounds, and in many cases on only one or two of these. The reason for obligate methylotrophy is a hereditary lack one or more enzymes, as will be discussed later.

Biochemically there are three pathways by which formaldehyde or CO_2 is converted into bacterial cell material (Fig. 3.1). We shall consider these first. The dissimilatory pathways by which the reduced C_1 growth substrates are oxidized to formaldehyde and CO_2 yielding energy to the cell are more diverse at the enzyme level, and will be considered later.

Assimilatory Pathways

Table 3.1 shows how the three assimilatory pathways are used by the various physiological groups of methylotrophs, and some examples are given of bacteria in each category which have been particularly closely studied. More complete lists of organisms using each biochemical pathway are to be found in the review of Colby *et al.* (1979).

Before we consider these pathways in detail, it is useful to know how they were originally unravelled. Three techniques have been used. The first,

Table 3.1 The Physiological and Biochemical Classification of Methylotrophic Bacteria

Abbreviations: MMA, monomethylamine; DMA, dimethylamine; TMA, trimethylamine; TMAO, trimethylamine N-oxide.

Physiological type	Carbon assimilation pathways	Examples	C_1 substrates supporting growth of each example
Heterotrophic methylotrophs	1. Serine pathway	*Pseudomonas* AM1	Methanol, MMA, formate
		Pseudomonas MA	MMA
		Methylobacterium organophilum	Methane, methanol
	2. Hexulose phosphate pathway	*Arthrobacter* P1	MMA, DMA, TMA
		Bacillus PM6	MMA, DMA, TMA, TMAO, tetramethylammonium
Autotrophic methylotrophs	Calvin cycle	Group 1. Phototrophs: *Rhodopseudomonas* spp.	Methanol, CO, formate
		Group 2. Chemoautotrophs:	
		Thiobacillus A2	Methanol, MMA, formate
		Paracoccus denitrificans	Methanol, MMA, formate
		Pseudomonas carboxydovorans	CO
		Group 3. *Pseudomonas oxalaticus*	Formate
Obligate methylotrophs	1. Serine pathway	*Methylomonas methanooxidans*	Methane, methanol only
		Methylosinus trichosporium	Methane, methanol only
	2. Hexulose phosphate pathway	*Methylomonas methanica*	Methane, methanol only
		Methylophilus methylotrophus	Methanol, MMA, DMA, TMA

and the simplest in principle is to give to whole cells growing on the substrate a relatively large dose of ^{14}C-labelled substrate. Cell samples are thereafter rapidly removed from the culture every few seconds and killed in hot ethanol. The aqueous ethanolic extract is concentrated and the radioactive compounds in it separated and identified. The percentage of the total radioactivity of a sample present in a particular compound at a given time is plotted as a function of time. In a sequence such as $A \rightarrow B \rightarrow C \rightarrow D$, the slope for the earliest product will be negative as radioactivity passes from it into other products, whose share of the radioactivity will increase with time (Fig. 3.2). If the first product can be identified in this way, it is usually possible to predict which enzymes might be involved in the conversion of A into B etc., and the presence of these enzymes can then be sought in crude cell-free extracts, and if necessary be purified and characterized. With facultative methylotrophs it is essential to compare the specific activity of these key enzymes with their specific activity in cells grown on a multicarbon substrate. If the enzymes play a role in the methylotrophic growth, they will probably be absent or present in only very low activity in cells grown heterotrophically. When the pathway has been identified in this way, mutants can then be sought that lack particular enzymes in the pathway. Such mutants

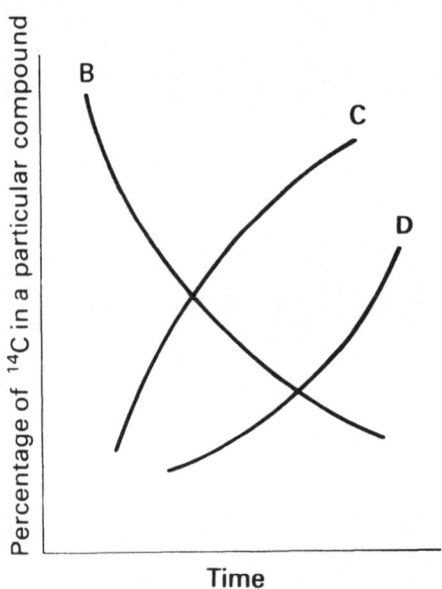

Fig. 3.2 Distribution of radioactivity in cell metabolites after ^{14}C-labelled A has been administered to growing cells. After separation of the radioactive products (the remaining quantity of labelled precursor A having first been removed), the percentage of radioactivity in each compound B, C, D, etc. as a function of the total radioactivity in that sample is plotted against the time of sampling. The first product to be labelled (B) will have a negative slope. Subsequent products will have a positive slope (C, D).

will have lost their ability to grow on one or more of the C_1 substrates (depending upon precisely which enzyme has been lost). In many cases of organisms that have been assigned to a particular carbon assimilation route, all three types of evidence have not been obtained. Often conclusions as to which pathway operates have been drawn only on the basis of the study of levels of activity of enzymes during methylotrophic and non-methylotrophic growth. Evidence of this type, unsupported by labelling or mutant evidence, is not always reliable, particularly when the enzyme activities measured are very low.

Most of the organisms listed in Table 3.1 have been thoroughly studied. Even so, evidence exists that some organisms may possess the enzymes of more than one pathway, although in many cases we have no evidence as to the extent to which alternative pathways may operate simultaneously. *Methylococcus capsulatus* (Bath strain) can produce enzymes of both the Calvin cycle and the hexulose phosphate pathway, and also can incorporate formate through enzymes of the serine pathway. *Arthrobacter* P1 during growth on methylamine uses enzymes of the hexulose phosphate pathway, but during heterotrophic growth on choline it uses enzymes of the serine pathway to assimilate glycine, while using the hexulose phosphate pathway to oxidize and assimilate formaldehyde from the methyl groups of choline. Thus the relationship between assimilation pathways is not necessarily mutually exculsive. We should note also that some enzymes of C_1-assimilatory pathways may have important roles in heterotrophic growth on certain substrates. Evidence based on measurements of enzyme activity in crude extracts together with slightly ambiguous results with labelling experiments in whole cells led for several years to the erroneous supposition that the hexulose phosphate pathway operated in the methylotrophic yeasts (see Chapter 4).

Carbon Assimilation via the Calvin Cycle. The Calvin cycle of carbon dioxide fixation, originally discovered in green plants, can also operate during non-photosynthetic growth in both chemoautotrophs and methylotrophs. The cycle is also sometimes known as the ribulose bisphosphate or reductive pentose-phosphate cycle. Without any experimental evidence, it was supposed until 1959 that all methylotrophs assimilated their carbon by this pathway, but in that year the occurrence of the serine pathway (see below) was demonstrated. The Calvin cycle is summarized in Fig. 3.3.

Details of the sugar-phosphate intermediates of the Calvin cycle will be found in any textbook of biochemistry. In most bacterial systems, whether autotrophic or methylotrophic, the electron donor for the cycle is probably NADH rather than NADPH. The overall stoicheiometry of the Calvin cycle is thus:

$$3CO_2 + 6NAD(P)H + 6H^+ + 9ATP \longrightarrow$$
$$\text{Glyceraldehyde 3-phosphate} + 6NAD(P)^+ + 9ADP + 8P_i \quad (3.1)$$

It is clear from this that assimilation of carbon at the level of CO_2 is much less efficient than its assimilation at a higher reduction level because of the heavy demands for ATP and reducing power.

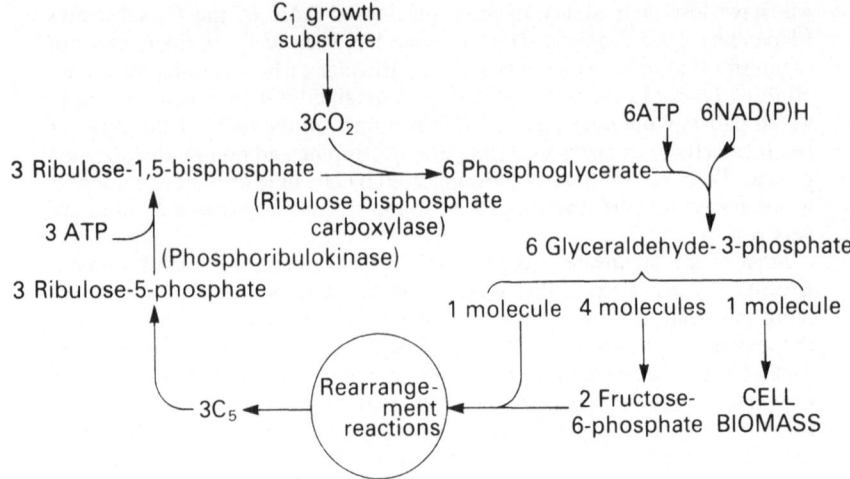

Fig. 3.3 The Calvin (ribulose 1,5-bisphosphate) cycle of carbon dioxide fixation.

The first two groups of autotrophic methylotrophs in Table 3.1 are really phototrophs and chemoautotrophs with an ability to use a limited number of reduced C_1 compounds. The *Rhodopseudomonas* species under photosynthetic conditions, i.e. anaerobically in the light, can assimilate reduced C_1 compounds instead of CO_2. A few of them can grow on methanol in the dark at low oxygen tensions. The bacteria in group 2 can all grow chemoautotrophically as well as methylotrophically. The third group of Calvin cycle bacteria has only one well characterized member, *Pseudomonas oxalaticus*, which has apparently no photosynthetic or chemoautotrophic capacities.

The key enzymes of the Calvin cycle are ribulose bisphosphate carboxylase (EC 4.1.1.39) (Equation 3.2)

$$CO_2 + \text{Ribulose 1,5-bisphosphate} \longrightarrow 2(3\text{-Phosphoglycerate}) \qquad (3.2)$$

and phosphophoribulokinase (EC 2.7.1.19) (Equation 3.3)

$$\text{Ribulose 5-phosphate} + ATP \longrightarrow$$
$$\text{Ribulose 1,5-bisphosphate} + ADP \qquad (3.3)$$

as indicated in Fig. 3.3. These have to be identified before the operation of the Calvin cycle can be substantiated.

Carbon Assimilation via the Hexulose Phosphate Cycle. Heterotrophic methylotrophs assimilate a substantial proportion of their growth substrate at the oxidation level of formaldehyde, and any CO_2 fixed (as in the serine pathway, see below and Fig. 3.1) is via reactions other than ribulose bisphosphate carboxylase. Two assimilation pathways have been recognized in bacteria and a third pathway has been found in methylotrophic yeasts (see

Chapter 4). These pathways also operate in obligate methylotrophs. The oxidation level of formaldehyde approximates to that of cell material, so that these pathways are energetically more economical than the Calvin cycle. The first of these pathways we will call the *hexulose phosphate cycle*. It is also called the ribulose monophosphate pathway of formaldehyde fixation. Other names may be found in the literature, including the allulose phosphate pathway, the ribose phosphate cycle and the Quayle cycle. The pathway is in some ways analogous to the Calvin cycle, but because formaldehyde is the reactant no reduction step is involved (Fig. 3.4).

The first *fixation* part of the cycle involves the two key reactions. These are the formation of D-*erythro*-L-*glycero*-3-hexulose 6-phosphate by condensation of formaldehyde with ribulose 5-phosphate (Equation 3.4), catalysed by *hexulose phosphate synthase* (EC 4.1.2.–) and the isomerization of the 3-hexulose phosphate to fructose 6-phosphate (Equation 3.5) catalysed by *hexulose phosphate isomerase*:

Ribulose 5-phosphate   3-Hexulose 6-phosphate   Fructose 6-phosphate

The second phase of the pathway (Fig. 3.4) involves *cleavage* of the hexose phosphate by two alternative routes. In one route fructose 6-phosphate is cleaved via fructose bisphosphate aldolase to dihydroxyacetone phosphate and glyceraldehyde 3-phosphate (as in the Embden–Meyerhof pathway of glycolysis). In the other, fructose 6-phosphate is converted via glucose 6-phosphate into 6-phosphogluconate, which is cleaved via 2-keto-3-deoxy-6-phosphogluconate to give glyceraldehyde 3-phosphate and pyruvate via enzymes of the Entner–Doudoroff pathway of carbohydrate breakdown.

The final stages of the pathway involve the *rearrangement* of triose- and hexose-phosphates to regenerate the ribulose 5-phosphate acceptor. In essence the process is $2C_6 + C_3 \rightarrow 3C_5$, but again there are two possible alternative pathways for these rearrangement reactions. These are shown in Fig. 3.5. One involves transaldolase, the other utilizes sedoheptulose bisphosphatase. Of the possible combinations from Figs. 3.4 and 3.5, i.e. (3.4)A + (3.5)(i), (3.4)B + (3.5)(i), (3.4)A + (3.5)(ii) and (3.4)B + (3.5)(ii), only the two combinations (3.4)A + (3.5)(ii) Embden–Meyerhof splitting + sedoheptulose bisphosphatase rearrangement and (3.4)B + (3.5)(i) Entner–Doudoroff splitting + transaldolase rearrangement, have so far been found.

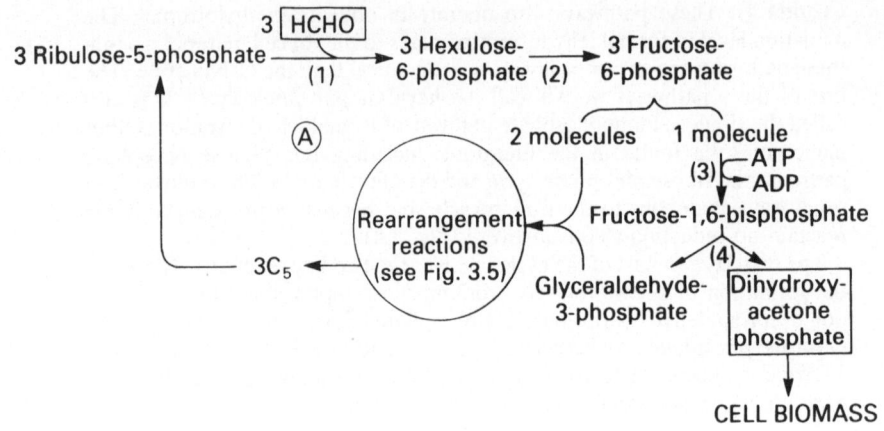

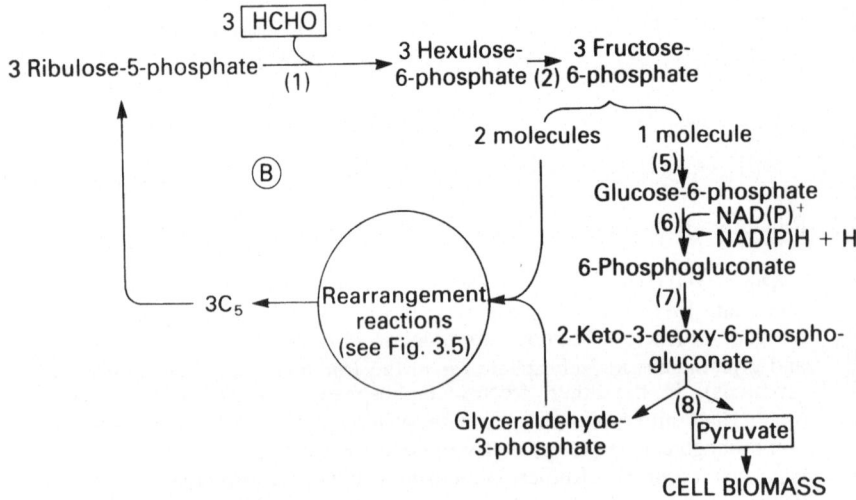

Fig. 3.4 The hexulose phosphate pathway of formaldehyde fixation showing the two alternative routes by which hexose molecules may be cleaved. A, the Embden–Meyerhof (fructose bisphosphate aldolase) variant; B, the Entner–Doudoroff (2-keto-3-deoxy-6-phosphogluconate aldolase) variant. The enzymes involved are: (1), hexulose phosphate synthase; (2), hexulose phosphate isomerase; (3), phosphofructokinase (EC 2.7.1.11); (4), fructose bisphosphate aldolase (EC 4.1.2.13); (5), hexose phosphate isomerase (EC 5.3.1.9); (6), glucose 6-phosphate dehydrogenase (EC 1.1.1.49); (7), 6-phosphogluconate dehydratase (EC 4.2.1.12); (8), 2-keto-3-deoxy-6-phosphogluconate aldolase (EC 4.1.2.14). The rearrangement reactions are shown in Fig. 35.

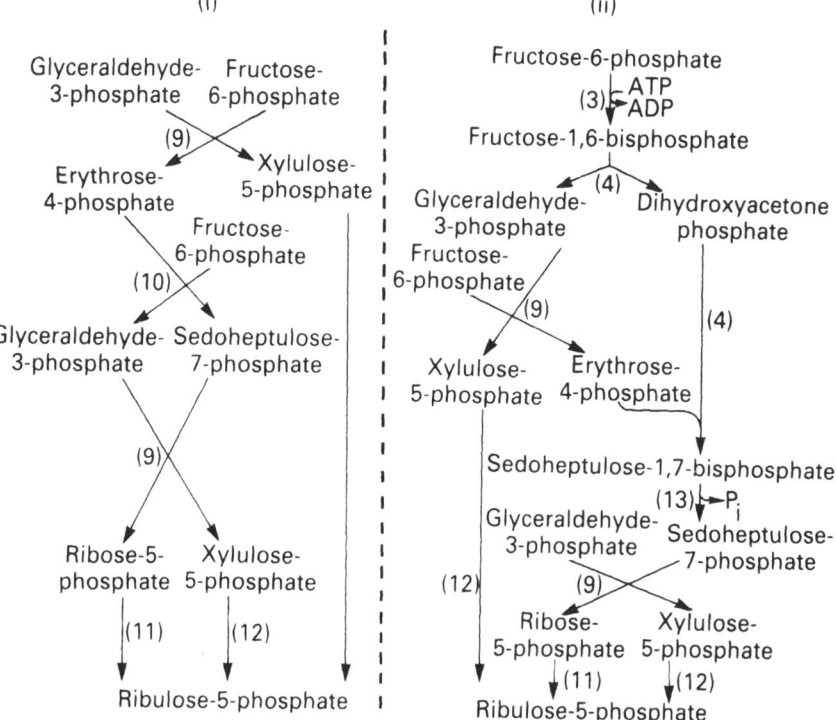

Fig. 3.5 Two alternative routes for the sugar phosphate rearrangement reactions of the hexulose phosphate cycle. Sequence (i) involves the enzyme (10), transaldolase (EC 2.2.1.2); and sequence (ii) involves (4), fructose bisphosphate aldolase; (13), sedoheptulose bisphosphatase (EC 3.1.3.37). Common to both alternatives are (9), transketolase (EC 2.2.1.1); (11), ribose phosphate isomerase (EC 5.3.1.6) and (12), ribulose phosphate 3-epimerase (EC 5.1.3.1).

There are resemblances between the hexulose phosphate cycle, the Calvin cycle and the dihydroxyacetone cycle (xylulose monophosphate pathway) of yeasts (see Chapter 4), and there is probably an evolutionary connection between the three. Some of the enzymes of the hexulose phosphate cycle also play an important role in formaldehyde oxidation in some organisms (see below).

Carbon Assimilation via the Serine Pathway. The serine pathway differs from the hexulose phosphate cycle in one important respect. Whereas in the latter cell carbon is assimilated only at the oxidation level of formaldehyde, in the serine pathway only 50 to 70% of the cell carbon arises from formaldehyde and the rest arises from CO_2. This fixation of CO_2, however, does not involve ribulose bisphosphate. The pathway gets its name because

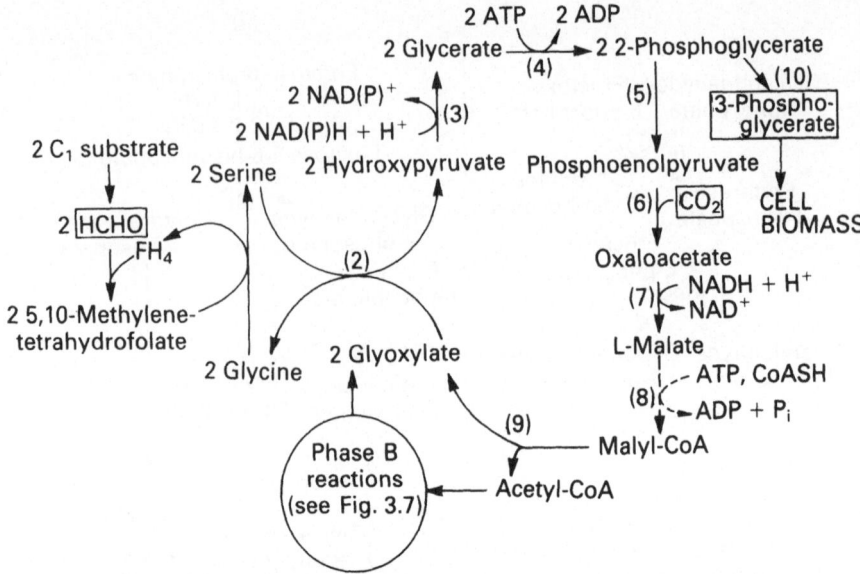

Fig. 3.6 The serine pathway. It can be divided into three phases: (A), conversion of formaldehyde and CO_2 (in boxes) into acetyl-coenzyme A; (B), conversion of acetyl-coenzyme A into glycine via glyoxylate; (C), net synthesis of a C_3 precursor of cell material (3-phosphoglycerate, in box) from glycine and formaldehyde. The overall reactions for each phase are given in the text (Equations 3.6–3.9). Not all organisms contain malyl-coenzyme A synthetase (reaction 8). The enzymes involved are as follows (1), serine hydroxymethyltransferase (EC 2.1.2.1); (2), serine-glyoxylate aminotransferase (EC 2.6.1.45); (3), hydroxypyruvate reductase (EC 1.1.1.81); (4), glycerate kinase (EC 2.7.1.31); (5), phosphoenolpyruvate hydratase (EC 4.2.1.11); (6), phosphoenolpyruvate carboxylase (EC 4.1.1.31); (7), malate dehydrogenase (EC 1.1.1.37); (8), malyl-coenzyme A synthetase (EC 6.2.1.9); (9), malyl-coenzyme A lyase (EC 4.1.3.24); (10), phosphoglycerate mutase (EC 2.7.5.3). FH_4 = tetrahydrofolate. For details of phase B see Fig. 3.7.

the first labelled product from [^{14}C]methanol (or formate, or methylamine) is serine. The pathway may be divided into three phases. In phase A (Fig. 3.6), a C_2-unit (acetyl-coenzyme A) is formed from formaldehyde and CO_2 (shown in boxes in Fig. 3.6), with the following stoicheiometry (Equation 3.6)

$$HCHO + CO_2 + 2\overset{*}{A}TP + 2NAD(P)H + 2H^+ + CoASH \longrightarrow$$
$$CH_3CO\text{—}SCoA + 2ADP + 2P_i + 2NAD(P)^+ \quad (3.6)$$

In phase B various reactions operate to convert the acetyl-coenzyme A via glyoxylate to glycine (Equation 3.7)

$$CH_3CO\text{—}SCoA + NAD^+ + 2H_2O \longrightarrow CHO.COOH +$$
$$CoASH + NADH + H^+ + 2H \quad (3.7)$$

and in the final phase (C), glycine and a further molecule of formaldehyde are converted into cell material via 3-phosphoglycerate (shown in a box in Fig. 3.6) (Equation 3.8)

$$HCHO + Glyoxylate + NAD(P)H + H^+ + ATP \longrightarrow$$
$$3\text{-Phosphoglycerate} + ADP + NAD(P)^+ \quad (3.8)$$

The serine pathway is found in large numbers of methylotrophs, both facultative and obligate (Table 3.1). The overall stoicheiometry of the cycle as seen from Fig. 3.6 is (Equation 3.9):

$$2HCHO + CO_2 + 3ATP + 2NAD(P)H + 2H^+ \longrightarrow$$
$$3\text{-Phosphoglycerate} + 2NAD(P)^+ + 2H + 3ADP + 2P_i + H_2O \quad (3.9)$$

The key enzymes of phase A of the cycle are serine hydroxymethyltransferase (EC 2.1.2.1), serine-glyoxylate aminotransferase (EC 2.6.1.45) which catalyses the conversion of serine to hydroxypyruvate simultaneously transferring the amino group to glyoxylate to give glycine, hydroxypyruvate reductase (EC 1.1.1.81) (Equation 3.10)

$$
\begin{array}{c}
CH_2OH \\
| \\
C{=}O \\
| \\
COOH
\end{array}
+ NAD(P)H + H^+ \rightleftharpoons
\begin{array}{c}
CH_2OH \\
| \\
CHOH \\
| \\
COOH
\end{array}
+ NAD(P)^+ \quad (3.10)
$$

malyl-coenzyme A synthetase (EC 6.2.1.9) (Equation 3.11)

$$L\text{-Malate} + ATP \rightarrow L\text{-Malyl-coenzyme A} + ADP + P_i \quad (3.11)$$

and malyl-coenzyme A lyase (EC 4.1.3.24) (Equation 3.12)

$$
\begin{array}{c}
CO.SCoA \\
| \\
CH_2 \\
| \\
CHOH \\
| \\
COOH
\end{array}
\longrightarrow
\begin{array}{c}
CO.SCoA \\
| \\
CH_3 \\
+ \\
CHO \\
| \\
COOH
\end{array}
\quad (3.12)
$$

Interestingly, malyl-coenzyme A synthetase is absent from *Pseudomonas* AM1 and several other serine pathway organisms. These bacteria must still be able to form malyl-coenzyme A, however, for if malyl-coenzyme A lyase is lost by mutation, the mutants fail to grow on C_1 substrates. It is important to note (Fig. 3.6) that the early stages of the serine pathway involve tetrahydrofolate. The precise details of the reaction of formaldehyde with tetrahydrofolate have not been established with certainty.

The second phase of the cycle (Fig. 3.7) involves the conversion of acetyl-coenzyme A to glyoxylate. Some organisms do this via isocitrate lyase (EC 4.1.3.1) and the enzymes of the tricarboxylic acid cycle. This is shown in Fig. 3.7a. Such organisms have increased activity of isocitrate lyase during methylotrophic growth. Bacteria of this class are sometimes called icl$^+$ methylotrophs. An example of such an organism is *Pseudomonas* MA (Table 3.1).

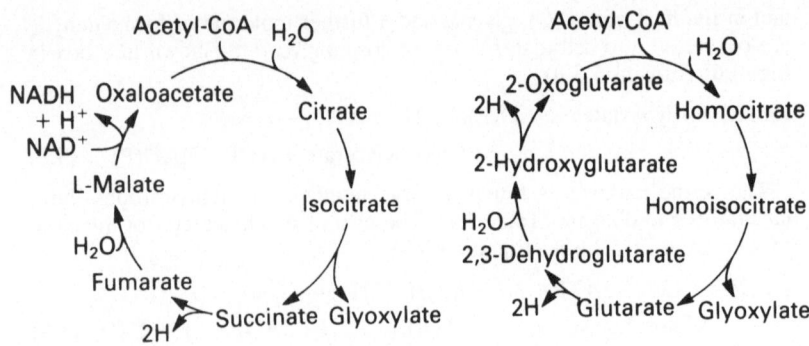

Fig. 3.7 Alternative mechanisms for phase B of the serine pathway (conversion of acetyl-coenzyme A into glyoxylate). (See Fig. 3.6.)

Many serine pathway methylotrophs, particularly a large group of pink-pigmented bacteria similar to *Pseudomonas* AM1 (Table 3.1), do not possess isocitrate lyase. Recent evidence has suggested a possible alternative route by which these organisms might convert acetyl-coenzyme A to glyoxylate. This involves homocitrate and homoisocitrate lyase (Fig. 3.7b). This alternative route involves a series of enzymes hitherto undescribed in bacteria (although some may be involved in lysine biosynthesis in yeasts). Further evidence is required, involving the characterization of the enzymes and obtaining mutants unable to synthesize the enzymes, before this pathway can be regarded as fully established. Organisms that lack isocitrate lyase for methylotrophic growth also do not use it for growth on acetate. It therefore seems likely that homoisocitrate lyase may also be used in these organisms for growth on acetate.

The CO_2-fixation reaction in the serine pathway involves the enzyme phosphoenolpyruvate carboxylase (EC 4.1.1.31) (Fig. 3.6).

Energy Requirement for the Three Assimilatory Pathways

As would be expected from the different reduction levels of their substrates, the amount of energy and reducing power required by the three assimilatory pathways is not the same. However we need to remember that in the case of the Calvin cycle, reducing power is generated during the oxidation of the reduced C_1 substrate to CO_2, so that the apparent high consumption of reducing power in Equation 3.1 does not actually apply in the few cases where the substrate is at the reduction level of formaldehyde or higher. In order to make a comparison between the pathways, Table 3.2 shows the net requirement for ATP when one molecule of pyruvate is synthesized via the three different bacterial assimilation pathways (using where necessary the glycolytic pathway to go from triose phosphate to pyruvate). The dihydroxyacetone pathway of yeast (Chapter 4) is also included for comparison. The hexulose phosphate and dihydroxyacetone pathways are the most efficient in terms of ATP consumption. whereas those pathways in which all or part

Table 3.2 Energy Balance for the Three Different Bacterial C₁-Assimilation Pathways

For comparison the dihydroxyacetone pathway of yeast (Chapter 4) is also given. The product is in each case pyruvate. FDP = fructose 1,6-bisphosphate; KDPG = 2-keto-3-deoxy-6-phosphogluconante.

Pathway	Route	Reactants	Molecules of ATP consumed
Calvin cycle		$3CO_2$	7
Hexulose phosphate cycle	FDP aldolase/sedoheptulose bisphosphatase	3HCHO	0
	KDPG aldolase/transaldolase	3HCHO	0
Serine pathway	Isocitrate lyase	$2HCHO + CO_2$	2
	Homoisocitrate lyase	$2HCHO + CO_2$	Not yet known
Dihydroxyacetone cycle (yeast)		3HCHO	1

of the cell carbon is assimilated as CO_2 have higher ATP requirements. This has important implications for the biotechnological use of methylotrophs, particularly for production of single-cell protein (Chapter 5).

Dissimilatory Pathways for Energy Generation

In this section we shall consider the pathways by which the various reduced C_1 compounds used as growth substrates are oxidized to CO_2, thereby making both energy and reducing power available to the cell. A general scheme showing these reactions is given in Fig. 3.8. One important feature of the pathways is that many of the dehydrogenases involved do not use nicotinamide nucleotides as electron acceptors. These are marked with a * in Fig. 3.8. The identity of the prosthetic groups and of the electron acceptors of some of these enzymes is not known with certainty. The enzymes have been assayed *in vitro* by the use of dyes, and it seems probable that the actual electron acceptors in the cell are more positive in electrode potential than nicotinamide nucleotides. The consequent implication of this is that less energy in the form of ATP can be obtained when these electron acceptors are reoxidized via the electron transport chain. In addition, several dissimilatory reactions consume rather than produce reducing power, because they use hydroxylase (mixed function oxidase, mono-oxygenase) mechanisms to oxidize methyl groups. These reactions are marked with a ⊕ in Fig. 3.8. An example of a mono-oxygenase reaction is Equation 3.13:

$$RH + O_2 + NAD(P)H + H^+ \longrightarrow ROH + NAD(P)^+ + H_2O \quad (3.13)$$

Pathways containing such an enzyme can only allow growth if the net production of reducing power exceeds the amount consumed by the mono-oxygenase reactions. It should be noted that most of the intermediates in Fig. 3.8 are themselves substrates for methylotrophic growth.

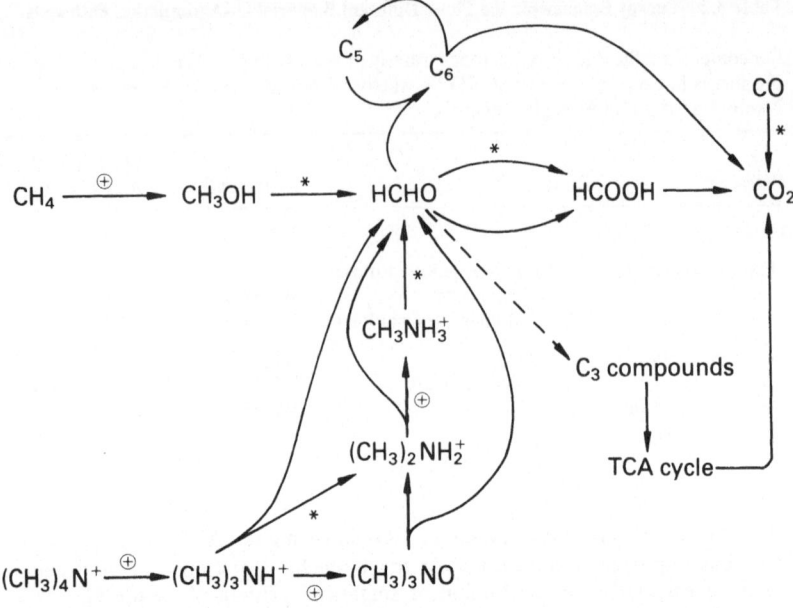

Fig. 3.8 Summary diagram of the oxidation pathways of reduced C_1 compounds. Intermediates in the pathway can usually also support methylotrophic growth. Dehydrogenases not using nicotinamide nucleotide coenzymes are marked *, and mono-oxygenase reactions are marked ⊕. Note the central role of formaldehyde.

Role of the Tricarboxylic Acid Cycle. As mentioned in Chapter 1, in many organisms the tricarboxylic acid cycle does not fulfil any important oxidative role during methylotrophic growth. This is proved by the fact that most obligate methylotrophs using the hexulose phosphate pathway of carbon assimilation do not seem to have a functional tricarboxylic acid cycle, since they lack the 2-oxoglutarate dehydrogenase system. Moreover their citrate synthase is not regulated by NADH. Mutants of the facultative methylotroph *Pseudomonas* AM1 (a serine pathway organism) that have lost 2-oxoglutarate dehydrogenase fail to grow on multicarbon compounds, and thus have become obligate methylotrophs. Revertants that regain the enzyme activity are again able to grow heterotrophically. Nevertheless, the tricarboxylic acid cycle enzymes between citrate and 2-oxoglutarate and between oxaloacetate and succinyl-coenzyme A fulfil an essential role in the synthesis of porphyrins and of amino acids of the glutamate family. Wild type organisms with the serine pathway (obligate as well as facultative) have low but measurable activities of 2-oxoglutarate dehydrogenase, and their citrate synthase is inhibited by 0.1 mM NADH. This may be a control mechanism to conserve carbon in the presence of an active tricarboxylic acid cycle.

In at least one organism, *Pseudomonas* MA (Table 3.1) the tricarboxylic acid cycle may play a role in energy generation. In this organism, a pathway has been proposed in which acetyl-coenzyme A from the first phase of the serine pathway is oxidized to CO_2 via the tricarboxylic acid cycle. Evidence supporting this hypothesis is the observation that phosphoenolpyruvate carboxylase of *Pseudomonas* MA is allosterically regulated by NADH and ADP, as might be expected if it is sited at a branch point between the assimilatory serine pathway and the proposed dissimilatory cycle.

Oxidation of Carbon Monoxide. Although carbon monoxide (CO) is a very toxic substance, not only to organisms with haemoglobin blood-pigments, but to all cells which contain cytochrome oxidase (cytochromes a/a_3), several bacteria can grow on it and obtain energy. Anaerobically it can be metabolized by methanogenic bacteria (Chapter 2) and can be used as a growth substrate in some cases. It can also be utilized as sole carbon source by various autotrophic methylotrophs. Photosynthetic *Rhodopseudomonas* species and some hydrogen bacteria can grow on CO. Some hydrogen bacteria have been given the name *carboxydobacteria* because of their ability to grow on CO oxidatively as sole carbon source. These organisms probably play an important ecological role in removal of excessive levels of the toxic pollutant CO from the atmosphere. *Pseudomonas carboxydovorans* (Table 3.1) has been the most extensively studied. In all carboxydobacteria, CO is oxidized via carbon monoxide dehydrogenase (Equation 3.14)

$$CO + H_2O + Acceptor \longrightarrow CO_2 + Reduced\ acceptor \qquad (3.14)$$

This enzyme is an iron–sulphur protein which also contains flavin-adenine dinucleotide (FAD) and molybdenum. It does not reduce nicotinamide nucleotides nor coenzyme F_{420}, and its natural electron acceptor has not yet been identified. It does not contain nickel as does the enzyme from some anaerobic bacteria. The energy obtained from this reaction is used by the carboxydobacteria to fix CO_2 via the Calvin cycle.

It is also known that CO is a substrate for methane mono-oxygenase (see below), and mono-oxygenation is a much more exergonic reaction than dehydrogenation (see Equation 3.15)

$$NADH + H^+ + CO + O_2 \longrightarrow CO_2 + H_2O + NAD^+ \qquad (3.15)$$

However, as yet no organism has been found whiich can grow on CO by this mechanism.

Oxidation of Cyanide. Although several actinomycetes and bacteria from both Gram-positive and Gram-negative genera have been isolated that are claimed to grow on cyanide, no reports of the enzymology of this have been made. One possible pathway would be via hydrolysis (Equation 3.16)

$$HCN + H_2O \longrightarrow HCONH_2 \xrightarrow{H_2O} NH_3 + HCOOH \longrightarrow$$
$$CO_2 + 2H^+ + 2e^- \qquad (3.16)$$

Oxidation of Methane, Methanol, Formaldehyde and Formate

These are successive intermediates in the oxidation of methane to CO_2, but each is also capable of sustaining growth of methylotrophic bacteria. The ability to use methane as growth substrate seems to be less widespread than the ability to grow on methanol. In the case of formaldehyde, because it is a very toxic substance, it can only serve as a growth substrate in continuous culture. In these conditions it has been found to be a good carbon and energy source.

Most methane-utilizing bacteria (methanotrophs) are *obligate*: they will only grow on methane or methanol, but recent work has identified some facultative methanotrophs that use the serine pathway for growth, for example *Methylobacterium organophilum* (Table 3.1). An important feature of methane-utilizing bacteria is the presence (as revealed by electron micrographs) of large quantities of intracellular membranes. These membranes may either be bundles of vesicular discs lying within the cell (Type I) or paired membranes lying round the cell periphery (Type II). These are illustrated in the review of Higgins (1979). Bacteria with Type I membrane systems utilize the hexulose phosphate cycle for growth while those with Type II membranes use the serine pathway. The facultative methanotrophs only contain their Type II membrane system when grown on methane. Cells grown on glucose or methanol have no membrane system. This suggests that the enzymes of methane oxidation may reside in the membranes, although recent work indicates that cells with few membranes may possess a soluble rather than a particulate methane mono-oxygenase system.

Oxidation of Methane. The methane molecule can only be attacked by a substitution mechanism. It was shown in 1970 that growth on methane is accompanied by the incorporation of an oxygen atom from gaseous oxygen into the molecule to give methanol. The enzymic mechanism of this reaction has been the subject of controversy. It appears that there are two different enzyme systems, one of which is particulate (i.e. associated with membranes in cell-free extracts), the other of which is soluble (i.e. cannot be sedimented within 2 h at a centrifugal force of $150,000 \times g$). These show a different sensitivity to inhibitors.

The soluble system was demonstrated in the thermophilic Bath strain of *Methylococcus capsulatus*, but may also occur in *Methylosinus trichosporium* (Table 3.1) along with the particulate mono-oxygenase system, the appearance of either pathway being a function of growth conditions. In the soluble system three proteins are involved: A, molecular weight 220 000 containing nonhaem iron and sulphur; B, molecular weight 15 000 with no detectable prosthetic group; and C, molecular weight 44 600 containing one molecular of FAD and one molecule of iron and sulphur per molecule. Protein C is reduced by NAD(P)H and protein A is the hydroxylase which seems only to bind the substrate when it has been reduced. The role of protein B has not yet been clarified (Equation 3.17).

$$
\begin{array}{c}
\text{NAD(P)H} + \text{H}^+ \diagdown \diagup \text{Protein C} \diagdown \diagup \text{Reduced} \diagdown \diagup \text{CH}_4 \\
\text{protein A} \\
\text{B} \\
\text{Reduced} \\
\text{NAD(P)}^+ \diagup \diagdown \text{protein C} \diagup \diagdown \text{Protein A} \diagup \diagdown \text{CH}_3\text{OH}
\end{array}
\qquad (3.17)
$$

The second system, purified from *Methylosinus trichosporium* was particulate, i.e. sedimented at $150000 \times g$ in the ultracentrifuge, but could be solubilized with phospholipase. Of the three components, the smallest, molecular weight 9400 has not been further characterized. The other two components were a soluble cytochrome c that could bind CO (molecular weight 13 000) and a colourless copper-containing protein molecular weight 47 000. With the purified enzyme, NAD(P)H could not act as electron donor. Ascorbate was the only active electron donor, presumably because it can reduce cytochrome c non-enzymically. In addition, methanol in the presence of purified methanol dehydrogenase could replace ascorbate as a reductant. Recent work has also identified the first (ABC) hydroxylation system in *Methylosinus trichosporium*. Moreover, the purified proteins of the *Methylococcus capsulatus* (Bath) system are functionally interchangeable with the corresponding proteins in *Methylosinus trichosporium*. Both enzyme systems are non-specific and will hydroxylate a very wide range of compounds other than methane and CO. Possible commercial applications of methane mono-oxygenase have been envisaged (Chapter 5). Other substrates include longer-chain alkanes, alkenes, dimethyl- and diethyl-ether, alicyclic, aromatic and heterocyclic compounds, and also ammonia which is hydroxylated to hydroxylamine (Equation 3.18)

$$NH_3 + NAD(P)H + H^+ + O_2 \longrightarrow NH_2OH + NAD(P)^+ + H_2O \quad (3.18)$$

Methanol is also a substrate, being converted to formaldehyde, but the K_m of the enzyme system for methanol is so high that methanol hydroxylation is probably not of any physiological significance in the cell.

The reaction sequence for the oxidation of methane to CO_2 is shown in Fig. 3.9.

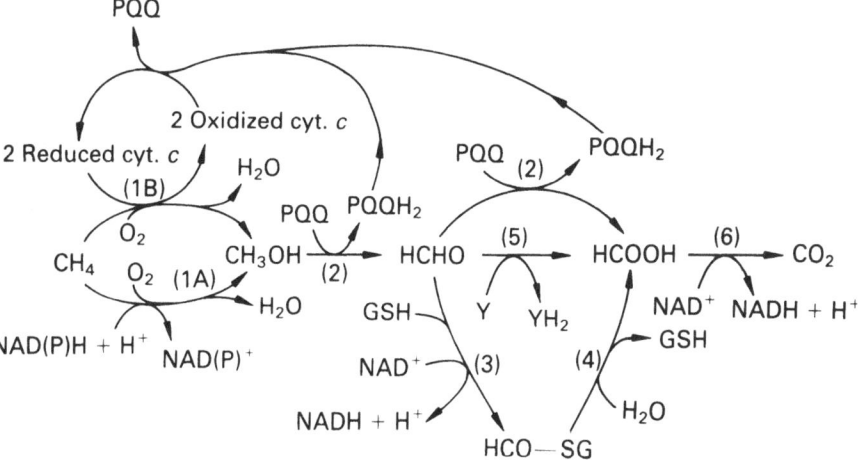

Fig. 3.9 Oxidation of methane and methanol by bacteria. The enzymes are (1A), methane mono-oxygenase system (NAD(P)H-dependent); (1B), methane mono-oxygenase (reduced cytochrome c-dependent); (2), methanol dehydrogenase (EC 1.1.99.8); (3), formaldehyde dehydrogenase (EC 1.2.1.1); (4), S-formyl-glutathione hydrolase (EC 3.1.2.12); (5) dye-linked aldehyde dehydrogenase (Y = dye); (6), formate dehydrogenase (EC 1.2.1.2). GSH is reduced glutathione, PQQ is pyrrolo-quinoline quinone (see text and Fig. 3.10).

Oxidation of Methanol. In contrast to yeasts (Chapter 4), bacteria oxidize methanol by means of a dehydrogenase (EC 1.1.99.8). Four types of methanol dehydrogenase have been identified, differing only slightly in substrate specificity and molecular weight. All have a spectral absorption maximum at 345 nm. Most methanol dehydrogenases show activity with alkan-1-ols with chain lengths up to C_{11}, and also with formaldehyde. When purified, they have a requirement for ammonia or a primary amine as activating agent. All contain a novel quinone prosthetic group called *methoxatin or pyrrolo-quinoline quinone* (PQQ) (Fig. 3.10). The prosthetic group occurs in a considerable number of other bacterial dehydrogenases, including glucose and ethanol dehydrogenases from *Acinetobacter calcoaceticus*. The name *quinoproteins* has been proposed for these enzymes. Methanol dehydrogenase was the first quinoprotein to be identified. The redox potential of the free $PQQ/PQQH_2$ couple is about $+120$ mV at pH 7.0, and methanol dehydrogenase is thought to react with the electron transport chain at the level of cytochrome c. Purified methanol dehydrogenase fails to react with cytochrome c, possibly because oxygen damages

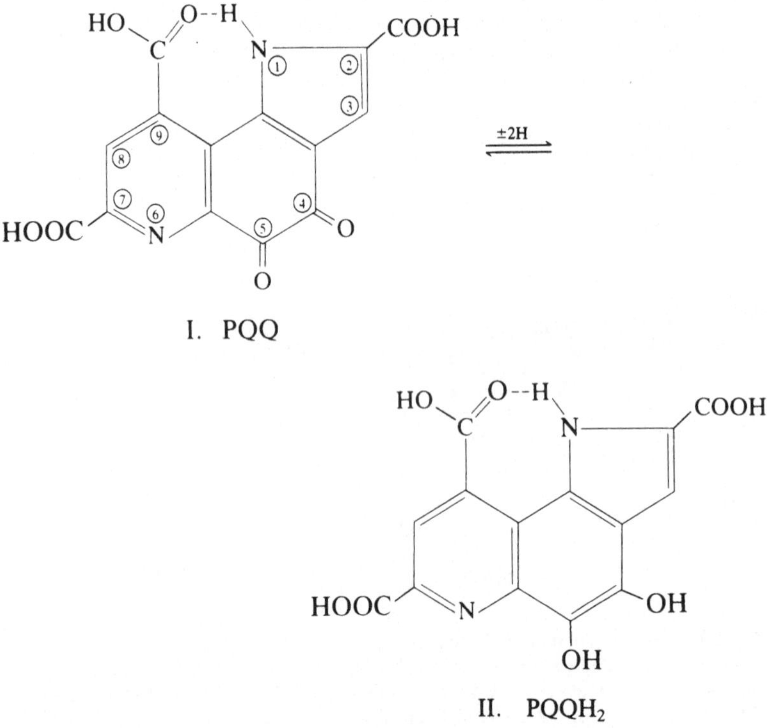

Fig. 3.10 Structure of the oxidized and reduced forms of methoxatin (pyrrolo-quinoline quinone, PQQ). The systematic name of I is 2,7,9-tricarboxy-1H-pyrrolo[2,3-f]quinoline-4,5-dione.

the functional coupling between the two in the cell. Using an anaerobic isolation procedure, methanol dehydrogenase preparations have been obtained that can reduce cytochrome c and which do not require the presence of ammonium as an activator. As soon as such preparations are exposed to oxygen, they lose the ability to reduce cytochrome c and require the presence of ammonium ions to reduce dyes. We may conclude that the mechanism of action resembles that of Equation 3.19

$$\text{CH}_3\text{OH} \searrow \quad \text{Enzyme-PQQ} \quad \nwarrow \nearrow 2 \text{ cyt } c_{red}$$
$$\text{HCHO} \nearrow \searrow \text{Enzyme-PQQH}_2 \nearrow \searrow 2 \text{ cyt } c_{ox} \tag{3.19}$$

The quinoprotein glucose dehydrogenase (EC 1.1.99.17) is important because it can be reversibly split into apoenzyme and PQQ and the apoenzyme can then be used as a biological assay for PQQ. With methanol dehydrogenase, removal of PQQ inactivates the apoenzyme. There is evidence (see below) that methylamine dehydrogenase may also be a quinoprotein.

Oxidation of Formaldehyde. Unlike methane and methanol, which are metabolized by only a few micro-organisms, almost all living cells contain dehydrogenases for formaldehyde and formate, because these two compounds are normal metabolites. Several different enzymes that catalyse formaldehyde oxidation occur in methylotrophs. The best characterized are (a) methanol dehydrogenase, which can oxidize formaldehyde as well as primary alcohols; (b) formaldehyde dehydrogenase (EC 1.2.1.1), which is NAD-dependent and for which the actual substrate is S-hydroxy-methylglutathione (Fig. 3.9, reaction 3; cf. Equation 4.5 in Chapter 4), and the product is S-formylglutathione. This enzyme thus requires the presence of a second enzyme, S-formylglutathione hydrolase to convert the product into formate (Fig. 3.9, reaction 4); (c) a dye-linked, non-specific aldehyde dehydrogenase (Fig. 3.9, reaction 5), which in some organisms seems to be the only enzyme present that can oxidize formaldehyde. This enzyme has recently been purified and shown to be a haemoprotein.

The reason for the great interest in the enzymes of methanol- and formaldehyde-oxidation is that the amount of energy available from electron transport and hence the molar growth yield of the cell will be lower if a non-nicotinamide nucleotide-dependent dehydrogenase is involved, in comparison with an NAD(P)-dependent system. This has biotechnological implications (Chapter 5).

In many organisms using the hexulose phosphate cycle of carbon assimilation, very low activities (or even none at all) have been detected for formaldehyde and formate dehydrogenases. It has been suggested that these organisms oxidize formaldehyde via the hexulose phosphate pathway and 6-phosphogluconate dehydrogenase (Fig. 3.11). The enzymes of this *dissimilatory hexulose phosphate cycle* (particularly glucose 6-phosphate and 6-phosphogluconate dehydrogenases) are very active in bacteria using the assimilatory hexulose phosphate pathway for growth on methanol or methylated amines. Bacteria of this group growing on methane have only low

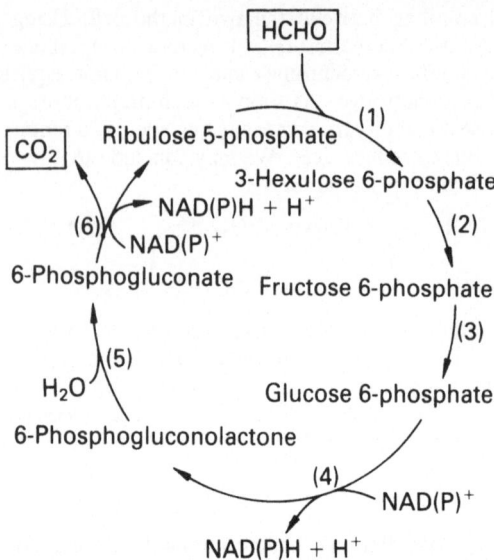

Fig. 3.11 The dissimilatory hexulose phosphate cycle. The overall reaction is given in the text (Equation 3.20). The enzymes are (1), hexulose phosphate synthase (EC 4.1.2.–); (2), hexulose phosphate isomerase; (3), hexose phosphate isomerase (EC 5.3.1.9); (4), glucose 6-phosphate dehydrogenase (EC 1.1.1.49); (5), 6-phosphogluconolactonase (EC 3.1.1.31); (6), 6-phosphogluconate dehydrogenase (EC 1.1.1.44).

activities of glucose 6-phosphate and 6-phosphogluconate dehydrogenases, and may oxidize methane directly via formate as shown in Fig 3.9.

The stoicheiometry of the dissimilatory hexulose phosphate cycle is as shown in Equation 3.20

$$\text{HCHO} + 2\text{NAD(P)}^+ + \text{H}_2\text{O} \longrightarrow \text{CO}_2 + 2\text{NAD(P)H} + 2\text{H}^+ \qquad (3.20)$$

It should be noted that *formate is not involved* as an intermediate. In *Methylophilus methylotrophus* the glucose 6-phosphate dehydrogenase (Fig. 3.11, reaction 4) is active with both NAD^+ and NADP^+, whereas two 6-phosphogluconate dehydrogenases are present, one specific for each nicotinamide nucleotide. The role of these two enzymes has not yet been clarified. *Pseudomonas* C in contrast contains only a single 6-phosphogluconate dehydrogenase active with both coenzymes. Lactonase activity (reaction 5 of Fig. 3.11) has not yet been demonstrated in these bacteria, but the reaction can also proceed non-enzymically. The activity of enzymes of the dissimilatory cycle seems to be modulated by several metabolites. This would be expected since it represents an important branch point, particularly in bacteria like *Methylophilus methylotrophus* that possess the Entner–Doudoroff variant of the assimilatory cycle. The importance of the dissimilatory hexulose phosphate cycle in the physiology of the cell awaits confirmation by characterization of the enzymes involved.

Oxidation of Formate. Most methylotrophs contain an NAD^+-dependent formate dehydrogenase (Fig. 3.9, reaction 6) and it has tended to be assumed that the enzyme is the same in all the organisms which have been examined. This may not be true. A soluble formate dehydrogenase has been purified to homogeneity from the autotrophic methylotroph *Pseudomonas oxalaticus* (Table 3.1). Its oxidation product is free CO_2, not bicarbonate, and CO_2 can be reduced to formate in the presence of NADH and purified enzyme. The enzyme contains 2 mol of flavin mononucleotide (FMN) per mol and also contains nonhaem iron and acid-labile sulphur. It is very unstable, and inactivated by oxygen. The enzyme seems not to have been purified from other methylotrophs. *P. oxalaticus* contains a second, non-NAD^+-dependent formate dehydrogenase, which is membrane-bound.

When formate functions as a growth substrate, it is oxidized to CO_2 to give energy. In *P. oxalaticus* the energy and reducing power so generated are used to fix CO_2 via the Calvin cycle. This is exceptional. In most other organisms that can grow on formate, the energy and reducing power are used to convert formate (via formyltetrahydrofolate) to 5,10-methylenetetrahydrofolate, which then enters the serine pathway.

Oxidation of Methylated Amines

The problem that confronts bacteria growing on methylated amines as sole carbon source is how to convert *N*-methyl groups into formaldehyde (Equation 3.21)

$$(CH_3)_4N^+ \xrightarrow{\quad} (CH_3)_3NH^+ \xrightarrow{\quad} (CH_3)_2NH_2^+ \xrightarrow{\quad}$$
$$\qquad HCHO \qquad\qquad HCHO \qquad\qquad HCHO$$

$$CH_3NH_3^+ \xrightarrow{\quad} NH_4^+ \qquad (3.21)$$
$$\qquad HCHO$$

Two different types of pathway for this conversion are known. One pathway involves dehydrogenases, and is thus potentially capable of operating under anaerobic conditions. The other involves mono-oxygenases, and is thus exclusively aerobic. The mechanisms are summarized in Figs. 3.12 and 3.13, and the test will refer to the enzymes on these digrams. For further details, see the review by Large (1981).

Tetramethylammonium Salts. The initial attack on tetramethylammonium salts is exclusively aerobic. Only a few bacteria can utilize quaternary ammonium compounds, and only one has been investigated biochemically. The enzyme consists of two proteins, uses NADPH or NADH as electron donor, and is not inhibited by CO (Fig. 3.12, reaction 1). This means that it is probably not a haemoprotein.

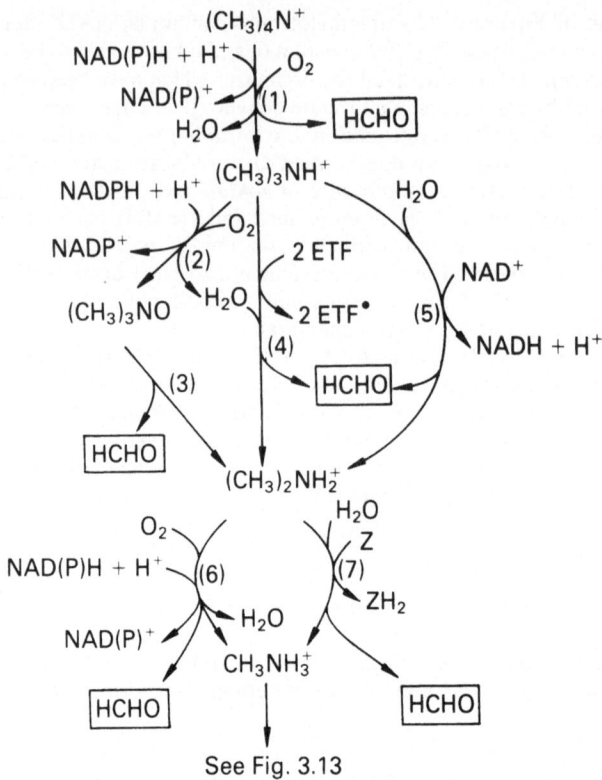

See Fig. 3.13

Fig. 3.12 Enzymes involved in the conversion of tetramethylammonium salts via trimethylamine and dimethylamine to monomethylamine. (1) Tetramethylammonium mono-oxygenase; (2), trimethylamine mono-oxygenase; (3) trimethylamine N-oxide aldolase (EC 4.1.2.32); (4), trimethylamine dehydrogenase (EC 1.5.99.7); (5), trimethylamine dehydrogenase (NAD-linked); (6), dimethylamine mono-oxygenase; (7), dimethylamine dehydrogenase (EC 1.5.99.–). Abbreviations: ETF electron-transferring flavoprotein (oxidized form), ETF⁺ electron-transferring flavoprotein (semiquinone half-reduced form); Z, unidentified electron acceptor.

Trimethylamine. Three different enzymes oxidizing trimethylamine have been identified (Fig. 3.12, reactions 2, 4 and 5). Two are dehydrogenases leading directly to the formation of dimethylamine and formaldehyde. One is NAD^+-dependent, but so far has been found only in one organism. The other is a flavoprotein, independent of nicotinamide nucleotides, which occurs in several organisms. It also contains nonhaem iron and sulphur. The flavin moiety is 6-S-cysteinylflavin-adenine dinucleotide, i.e. FAD covalently linked to the polypeptide of the apoenzyme. The natural electron acceptor of this enzyme is not a cytochrome but a second flavoprotein, an *electron-transferring flavoprotein* (denoted by ETF in Fig. 3.12) which is a single-electron acceptor. This probably then reacts with the electron transport chain at the level of cytochrome b.

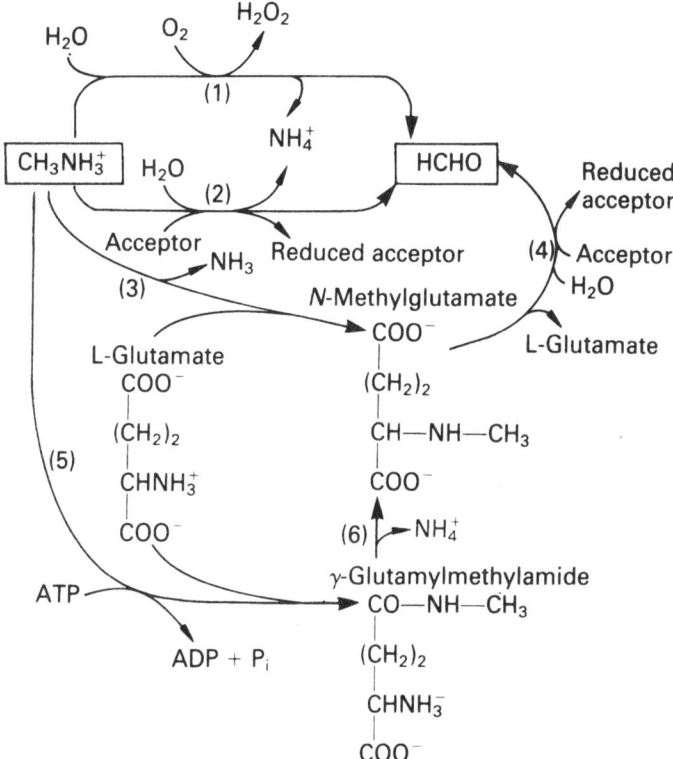

Fig. 3.13 Enzymic pathways of methylamine oxidation. The enzymes are: (1), methylamine oxidase; (2), methylamine dehydrogenase (EC 1.4.99.3); (3), N-methylglutamate synthase (EC 2.1.1.21); (4), N-methylglutamate dehydrogenase (EC 1.5.99.5); (5), γ-glutamylmethylamide synthetase (EC 6.3.4.12); (6), enzyme not characterized.

The other enzyme catalysing trimethylamine oxidation is a mono-oxygenase, specific for NADPH which converts trimethylamine into trimethylamine N-oxide. It may be a flavoprotein. The N-oxide is converted non-oxidatively into dimethylamine and formaldehyde by a further enzyme, trimethylamine N-oxide aldolase (Fig. 3.12, reaction 3). Different forms of this enzyme have been found in Gram-positive and Gram-negative bacteria.

Dimethylamine. With this substrate also, both a mono-oxygenase and a dehydrogenase are known (Fig. 3.12, reactions 6 and 7). Both lead to the formation of dimethylamine and formaldehyde. In most bacteria dimethylamine is oxidized by the mono-oxygenase, a very unstable enzyme which contains cytochrome P-420 and which is extremely sensitive to CO. In aerobically grown *Hyphomicrobium* X dimethylamine mono-oxygenase is the usual catabolic enzyme, but under anaerobic conditions, when the

47

organism can grow on nitrate in place of oxygen as terminal electron acceptor, it contains a dimethylamine dehydrogenase with similar properties to trimethylamine dehydrogenase. The two enzymes, however, can be separated easily.

Methylamine. The bacterial oxidation of methylamine is even more complicated because three separate oxidation pathways have been found (Fig. 3.13), and in many cases it is not known whether more than one may occur in the same organism.

The first mechanism is oxidation via an amine oxidase with the formation of ammonia, formaldehyde and hydrogen peroxide (Fig. 3.13, reaction 1). This pathway has so far been identified only in the genus *Arthrobacter*. The enzyme has been only partially characterized but seems to resemble the copper-containing amine oxidases (EC 1.4.3.6).

Many bacteria from all of the different categories in Table 3.1 contain methylamine dehydrogenase (Fig. 3.13, reaction 2). This enzyme is unusually heat-stable. It has been purified from several different bacterial strains. The prosthetic group, once thought to be pyridoxal phosphate, is now believed to be PQQ, probably covalently linked to the protein. The electron acceptor *in vivo* is thought to be a blue copper protein called amicyanin, which is reoxidized by cytochrome *c*.

The third mechanism for methylamine oxidation is indirect, involving the formation and subsequent reoxidation of *N*-methyl amino acids (Fig. 3.13, reactions 3, 4, 5 and 6). In both these two systems glutamate plays a catalytic role and methylamine is converted to formaldehyde and ammonia. The synthase (reaction 3) contains FMN as prosthetic group (although it does not catalyse an oxidation) while *N*-methylglutamate dehydrogenase contains FAD. This is not nicotinamide nucleotide-linked, although an NAD^+-linked *N*-methylglutamate dehydrogenase has also been reported. γ-Glutamyl-methylamide (Fig. 3.13) probably plays a similar role to *N*-methyl-glutamate, although the enzyme that is thought to catalyse reaction 6 in Fig. 3.13 has not been characterized.

Oxidation of Methylated Sulphur Compounds

There remains a great deal of work to be done in this area. Only two compounds have been studied in detail, and each in only a single organism. The two compounds are trimethylsulphonium chloride and dimethyl sulphoxide.

Pseudomonas MS will grow on trimethylsulphonium salts, with the transfer of one methyl group to tetrahydrofolate (Fig. 3.14). The cells grow by oxidation of this one methyl group, and the dimethyl sulphide also formed is not metabolized further. This organism is the only case so far discovered where tetrahydrofolate derivatives are involved in the *oxidation* of C_1 compounds as well as in their assimilation (Fig. 3.14), although such a role has been postulated in methanol oxidation. *Pseudomonas* MS uses the serine pathway of carbon assimilation involving isocitrate lyase.

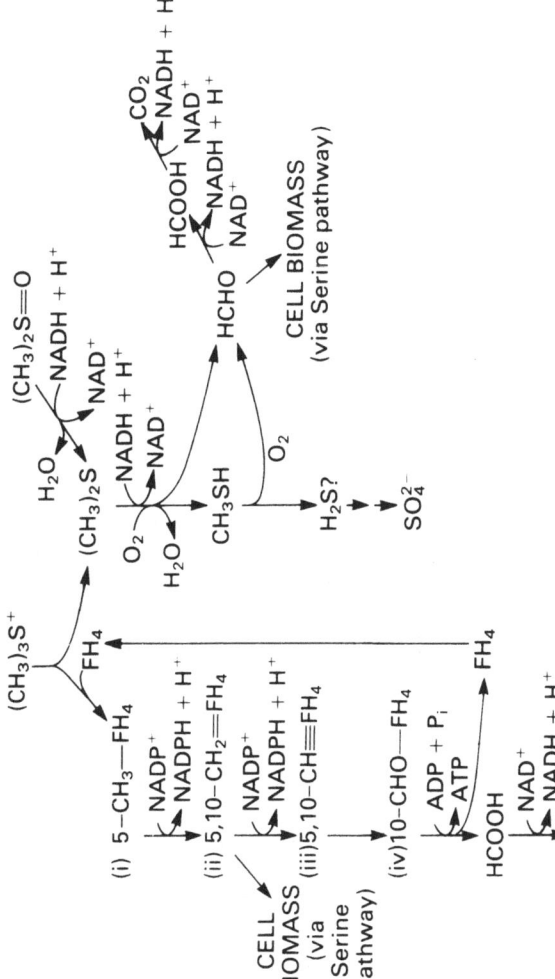

Fig. 3.14 Carbon metabolism of methylated sulphur compounds. Not all the enzymes have been properly characterized, so individual names are not given. *Pseudomonas MS* metabolizes trimethylsulphonium salts as shown on the left, *Hyphomicrobium S* metabolizes dimethyl sulphoxide as shown on the right. FH₄ = tetrahydrofolate. The derivatives of tetrahydrofolate involved are (i) 5-methyltetrahydrofolate; (ii), 5,10-methylenetetrahydrofolate; (iii) 5, 10-methenyltetrahydrofolate; (iv), 10-formyltetrahydrofolate.

Recent work on the utilization of dimethyl sulphoxide by bacteria has resulted in the isolation of an organism *Hyphomicrobium* S that metabolizes this compound as sole carbon, energy and sulphur source. It is converted to sulphate via dimethyl sulphide as shown in Fig. 3.14. The enzymes involved have not yet been characterized.

Electron Transport in Methylotrophic Bacteria

No account of the pathways of dissimilation of reduced C_1 compounds is complete without some information on how energy is obtained. This requires three kinds of information. The first is knowledge of the nature and order of the redox components of the electron-transport pathways in methylotrophs. The second is knowledge of the level at which electrons obtained by oxidation of the various reduced C_1 substrates enter the electron-transport chain, because unlike mammalian mitochondrial dehydrogenases, most of the dehydrogenases we have discussed do not generate reduced nicotinamide nucleotides. The third thing we need to know is the number of molecules of ATP that can be formed per pair of electrons by a particular electron-transport pathway in defined growth conditions. Much progress has been made in these fields in the last 10 years.

The first important point is that the electron-transport chain during methylotrophic growth may be quite different in composition and function to the chain in the same organism growing heterotrophically. Secondly, different methylotrophs differ in the composition of their electron-transport chains. Thirdly, the chain may alter according to the nature of the reduced C_1 substrate. This is particularly important during growth on substrates like methane or methylated amines, where an initial hydroxylation step may remove reducing power either as NAD(P)H or as reduced cytochrome *c*.

The Role of Cytochrome *c*. Cytochrome *c* seems to play an essential role in the oxidation of methanol (and probably also of methylated amines in some organisms), but in the heterotrophic methylotroph *Pseudomonas* AM1 for instance, it is not essential for growth on non-C_1 substrates in conditions of carbon excess. If the non-C_1 substrate is growth-limiting however, then cytochrome *c* is involved. The situation is complicated by the presence in most methylotrophs of two or more different *c*-type cytochromes, some soluble and some membrane-bound. Some of these are not typical basic proteins of the cytochrome *c*-type found in eukaryotes, and they have the rather unusual property of being able to react with CO. Whether this fact is of physiological importance to the cell is not yet known. Recent work has clearly shown that methanol dehydrogenase feeds its electrons into the electron transport chain at the level of cytochrome *c*, which may under intracellular conditions be the actual electron acceptor for the quinone prosthetic group of methanol dehydrogenase. Concentrations of cytochrome *c* are always elevated in cells grown on methanol compared with cells grown on non-C_1 substrates. Cytochrome *c* seems to play this key role in all organisms irrespective of the terminal oxidase, whether it be cytochrome a/a_3 or cytochrome o. Other C_1 substrates such as amines oxidized via dehydrogenases may feed their electrons into the chain at the level of cytochrome b (via a flavoprotein) or via a blue protein containing copper called amicyanin which then reduces cytochrome c.

The Electron-Transport Chain of *Pseudomonas* AM1. *Pseudomonas* AM1 is a well investigated pink-pigmented facultative heterotroph that uses the serine pathway of carbon assimilation during growth on methanol or formate (Table 3.1). In this bacterium the terminal oxidase seems always to be cytochrome a/a_3, but there are two b-type cytochromes. Under conditions of carbon excess, cytochrome c is probably not involved in electron transport from NADH, with a corresponding reduction in the ATP yield per pair of electrons, due to the loss of one of the proton-translocating loops (see below). There are at least two different cytochromes c in *Pseudomonas* AM1, and whether the two fulfil different roles in the pathway has not yet been established. Mutants lacking cytochrome c use a pathway in which electrons pass directly from cytochrome b to cytochrome a/a_3, with a lower yield of ATP. They are unable to grow on methanol or methylamine. During normal growth, studies of ATP synthesis coupled to electron transport in membrane vesicles suggest that oxidation of methanol to formaldehyde produces one ATP, whereas oxidation of succinate to fumarate produces two, and oxidation of NADH to NAD^+ produces three (probably only during carbon-limited growth conditions, otherwise two). Possible proton-translocating loops are thought to lie between the cytochromes b and cytochrome c, between methanol dehydrogenase and cyctochrome c and two between NADH and the b cytochromes (see Anthony, 1981).

The Electron-Transport Chain of Other Methylotrophs. There are considerable variations in the components of the electron-transport chain in different methylotrophs. The two organisms that have also been examined, *Methylophilus methylotrophus* (an obligate methylotroph, Table 3.1) and *Paracoccus denitrificans* (an autotrophic methylotroph, Table 3.1) both contain cytochrome o, which is a CO-reactive b-type cytochrome that can function as an alternative terminal oxidase. In the case of the latter organism, mutants lacking cytochrome c fail to grow on methanol, although they grow on all other substrates. The details of the cytochromes in these two organisms are discussed by Anthony (1981).

Recent work has shown that in both of these organisms methanol dehydrogenase and cytochrome c are both located on the periplasmic ('outside') surface of the cell membrane, from which methanol dehydrogenase is readily released. Cytochrome a/a_3 is sited on the cytoplasmic ('inside') surface of the membrane, and it has been shown that when a pair of electrons passes from cytochrome c to oxygen, there is a net inward translocation of $2e^-$ but no net proton translocation. This confirms that the maximum possible yield of ATP per mol of methanol oxidized to formaldehyde is 1 (Dawson and Jones, 1981*a*). It also implies that formaldehyde is produced outside the membrane (Fig. 3.15) and must pass through it by

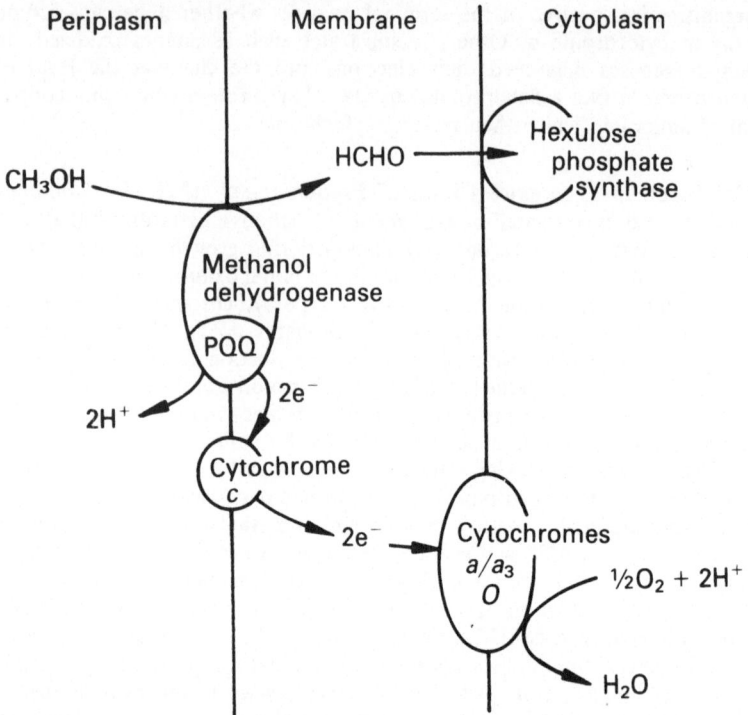

Fig. 3.15 A possible model for the redox arm organization of the methanol dehydrogenase system of *Methylophilus methylotrophus*, based on data of Dawson and Jones (1981a). PQQ = prosthetic group of methanol dehydrogenase (see text and Fig. 3.10).

simple or facilitated diffusion before becoming available for hexulose phosphate synthase (*M. methylotrophus*) or formaldehyde dehydrogenase (*P. denitrificans*). This single redox pathway through the membrane is called a redox arm (Dawson and Jones, 1981b).

Electron transport during growth on methane has been reviewed by Higgins (1981).

The results so far in this field only emphasize the need for more information. This is especially true for C_1 growth substrates other than methanol.

Control of Growth on C_1 Compounds

Two kinds of control mechanism exist in bacteria and we will discuss these separately.

Control at the Level of Enzyme Activity. By this is meant activation or inhibition of pre-existing enzymes by intracellular metabolites. Since the separation between assimilatory and dissimilatory routes in methylotrophs is sharp, control will only be necessary at those points in metabolic pathways where the two routes diverge. At these points strict control will be necessary to direct metabolites in the required direction. In an autotrophic methylotroph assimilating carbon as CO_2, where the pathways of assimilation and dissimilation are totally separate, a different kind of control mechanism must exist from the situation in, for example, hexulose phosphate cycle bacteria.

In Calvin cycle bacteria, the activities of key enzymes of the cycle, phosphoribulokinase, ribulose bisphosphate carboxylase, phosphoglycerate kinase and fructose/sedoheptulose bisphosphatase are regulated by both energy charge and NADH concentration. Phosphoenolpyruvate and 6-phosphogluconate also affect these enzymes. Further details of control of this pathway are discussed in McFadden (1978).

In methylotrophs using the hexulose phosphate cycle, there are two possible branch points between assimilation and dissimilation. In organisms using the Embden–Meyerhof variant, the branch point is at fructose 6-phosphate, which may be phosphorylated (for assimilatory purposes, see Figs. 3.4(A) and 3.5(ii)) or converted to glucose 6-phosphate (for oxidation via the dissimilatory hexulose phosphate cycle). In organisms using the Entner–Doudoroff variant, the branch point is at 6-phosphogluconate where the assimilatory (Fig. 3.4(B)) and dissimilatory (Fig. 3.11) cycles diverge. NAD(P)H and ATP inhibit the 6-phosphogluconate dehydrogenases of M. methylotrophus and Pseudomonas C, so that when the levels of ATP and reducing power in the cell are high, carbon is directed into the assimilatory pathway.

One of the key problems in the study of the serine pathway is that of how carbon at the level of formaldehyde is directed either via serine hydroxymethyltransferase into the assimilatory pathway or via one of the enzymes oxidizing formaldehyde to CO_2. The answer to this is not as yet known. Some workers have suggested that tetrahydrofolate plays a role in this.

Control at the Level of Enzyme Synthesis. By this term we mean induction or repression of enzyme synthesis by the growth substrate or a metabolite thereof. In this section also we shall consider isofunctional enzymes, that is enzymes that catalyse the same reaction but which are induced or repressed by different growth substrates (and which may also be subject to control at the activity level by different metabolites).

In the Calvin cycle growth on multicarbon substrates represses the unique enzymes of the cycle, phosphoribulokinase, ribulose bisphosphate carboxylase and fructose/sedoheptulose bisphosphatase. The evidence does not suggest that repression is co-ordinate.

In the case of the hexulose phosphate cycle, many of the bacteria which use this pathway are obligate methylotrophs, and so we know little as yet about the effect on the synthesis of enzymes of the cycle of multicarbon growth substrates.

The serine pathway is one of the few pathways where isofunctional enzymes have been found. Such enzymes differ in properties as proteins, but catalyse the same reaction. Only one of a pair of such enzymes is specifically induced by growth on reduced C_1 compounds. Examples include serine hydroxymethyltransferase and phosphoenolpyruvate carboxylase in *Methylobacterium organophilum*, and phosphoenolpyruvate carboxylase (probably a metabolic branch point, see above) and isocitrate lyase in *Pseudomonas* MA. These enzymes represent a means whereby an enzyme that functions in a methylotrophic pathway, but may have other functions in the cell, can be controlled by independent mechanisms. The control by repression of levels of methylotrophic enzymes in facultative methylotrophs using the serine pathway is discussed by O'Connor (1981).

As far as the enzymes of energy generation (dissimilation) are concerned, these are in most cases partially or completely repressed during growth on non-methylotrophic substrates.

The fact that various combinations of the different assimilatory and dissimilatory pathways occur in different organisms means that it is rather unlikely that the genes for the assimilatory and dissimilatory enzymes are part of a single unit of genetic control (an operon). The only evidence that might support this idea of a single operon is the isolation of mutants of *Pseudomonas aminovorans* (a serine pathway facultative methylotroph) which lack not only the enzymes for the oxidation of di- and tri-methylamine, but also various key assimilatory enzymes. Such an observation could also however be explained by the occurrence of the genes for the essential enzymes of methylotrophic growth on a plasmid, but as yet no evidence for this has been obtained.

The topic of control of growth on C_1 compounds is discussed by Quayle (1980). We need to know much more about the control of methylotrophic enzymes in order to establish whether genetic engineering can be useful to us by enabling us to make alterations in the properties of methylotrophs, which might thereby become more suitable for various biotechnological applications of the kind described in Chapter 5.

Summary

Methylotrophs may be either obligate, i.e. unable to grow on any substrates except reduced C_1 compounds, or facultative. Facultative organisms have alternative modes of existence. If the alternative is only heterotrophic growth on multicarbon substrates, the organisms are called heterotrophic methylotrophs; whereas if the alternative can be chemoautotrophic growth at the expense of inorganic compounds, the organisms are autotrophic methylotrophs. The latter organisms use the Calvin cycle to assimilate their carbon, having obtained energy by the oxidation of the growth substrate to CO_2. Non-autotrophic methylotrophs grow on reduced C_1 compounds via pathways in which the substrate is incorporated into cellular carbon wholly or partially at the oxidation level of formaldehyde. There are two pathways that have been demonstrated in bacteria, the hexulose phosphate cycle and the serine pathway. The former, in which all the cell carbon is derived from formaldehyde, is significantly less energy-demanding than the serine pathway and is used by most (but not all) obligate methylotrophs. The serine pathway involves 30 to 50% of the cellular carbon arising from CO_2, the remainder arising from formaldehyde. Finding out which of the three pathways operates in a particular species involves

1. examination of the early fixation products of ^{14}C-labelled growth substrates.
2. demonstration of the presence of key enzymes of the relevant pathway and the absence of enzymes of the other C_1 assimilation pathways, and
3. obtaining mutants unable to grow on reduced C_1 compounds and locating the position of the enzymic block.

It has been shown in some cases that enzymes of more than one assimilation pathway may be present during growth on certain substrates.

Aerobic methylotrophs obtain their energy for growth exclusively by the oxidation of their growth substrate to CO_2^-. In most cases (although some exceptions are known) the oxidation pathways do not involve the tricarboxylic acid cycle. Indeed in some organisms using the hexulose phosphate pathway of carbon assimilation, a key enzyme of the cycle, 2-oxoglutarate dehydrogenase is absent. This confers the phenotype of obligate methylotrophy. Some organisms (particularly those using the Calvin cycle and the serine pathway) have linear sequences of energy generation of the type $CH_4 \rightarrow CH_3OH \rightarrow HCHO \rightarrow HCOOH \rightarrow CO_2$ (Fig. 3.9), and $(CH_3)_nN \rightarrow HCHO$, etc. (Figs. 3.12 and 3.13), whereas many (if not all) organisms with the hexulose phosphate carbon assimilation pathway oxidize their growth substrate via the dissimilatory hexulose phosphate cycle $HCHO + C_5 \rightarrow C_6 \rightarrow C_5 + CO_2$ (Fig. 3.11). The detailed enzymology of the reactions for the oxidation of methane, methanol, formaldehyde, formate, methylated amines, methylated sulphur compounds and carbon monoxide is described. Many of the enzymes involved do not use NAD^+ or $NADP^+$ as electron acceptors, but acceptors whose redox potentials are probably higher. Thus the energy potentially available as ATP is less. The enzymes are of great intrinsic interest because they contain some novel prosthetic groups: flavins linked covalently via unusual bonds and novel quinones (methoxatin or PQQ).

What is known of the mechanisms of electron transport and ATP formation is discussed with particular reference to the oxidation of methanol. The mechanism of methanol oxidation is unusual because the high potential (about $+120\,mV$) of the $PQQ/PQQH_2$ couple means that electrons are fed into the electron transport chain at the level of cytochrome c, so that only one ATP can be formed per mol of methanol oxidized to formaldehyde. Unusual multiple c-type cytochromes seem to be present in all methylotrophs so far investigated. The role of these has not yet been completely established.

Because of the unique nature of many of the enzymes involved, strict repressive control of assimilatory and dissimilatory enzymes in facultative methylotrophs would be expected. This is indeed the case, and a few instances have been documented of unique 'methylotrophic' isofunctional enzymes to fulfil this need for control. The other crucial control mechanism must lie at the level of enzymic activity, to determine whether carbon at the level of formaldehyde is channelled into the oxidative or the dissimilatory pathway. A little is known about this in organisms using the hexulose phosphate cycle for growth, but virtually nothing about how it occurs in the serine pathway. There is a need for further information on control of methylotrophic metabolism to enable us to assess the potential of genetic engineering to improve methylotrophic strains for particular biotechnological applications.

References

COLBY, J., DALTON, H. and WHITTENBURY, R. (1979). Biological and biochemical aspects of microbial growth on C_1 compounds. *Annual Review of Microbiology* **33**, 481–517.

DAWSON, M.J. and JONES, C.W. (1981a). Energy conservation in the terminal region of the respiratory chain of the methylotrophic bacterium *Methylophilus methylotrophus*. *European Journal of Biochemistry* **118**, 113–118.

HIGGINS, I.J. (1979). Microbial biochemistry of methane. Part 2. Methanotrophy. In *Microbial Biochemistry* (International Review of Biochemistry, vol. 21), pp. 300–353. Edited by J.R. Quayle. University Park Press, Baltimore.

HIGGINS, I.J. (1981). Respiration in methylotrophic bacteria. In *Diversity of Bacterial Respiratory Systems*, vol. 1, pp. 187–221. Edited by C.J. Knowles. CRC Press, Boca Raton, Florida.

McFADDEN, B.A. (1978). Assimilation of one-carbon compounds. In *The Bacteria*, vol. 6. pp. 219–304. Edited by L.N. Ornston and J.R. Sokatch. Academic Press, New York.

QUAYLE, J.R. (1980). Aspects of the regulation of methylotrophic metabolism. *FEBS Letters* **117** (Suppl.) K16–K27.

The following references are all to chapters in *Microbial Growth on C_1 Compounds: Proceedings of Third International Symposium*. Edited by H. Dalton. Heyden & Son, London.

ANTHONY, C. (1981). Electron transport in methylotrophic bacteria, pp. 220–230.

DAWSON, M.J. and JONES, C.W. (1981b). Chemiosmotic aspects of respiratory chain energy conservation in *Methylophilus methylotrophus*, pp. 251–257.

LARGE, P.J. (1981). Microbial growth on methylated amines, pp. 55–69.

O'CONNOR, M.L. (1981). Regulation and genetics in facultative methylotrophic bacteria, pp. 294–300.

ZATMAN, L.J. (1981). A search for patterns in methylotrophic pathways, pp. 42–54.

4 Physiology and Biochemistry of Methylotrophic Yeasts

This chapter will be confined to a discussion of methylotrophic yeasts. Although certain mycelial fungi have been reported to grow on methanol, they have not been extensively studied. The reason for considering yeasts in a separate chapter from bacteria is that most of the key reactions of the assimilatory and dissimilatory pathways for C_1 compounds are different. This is presumably because eukaryotes appeared much later in the history of living things than did prokaryotes. Some yeasts have been reported to grow slowly on methane and methylamine, but these capabilities are relatively rare and very little is known about them. Accordingly the remainder of this chapter will discuss the growth of yeasts on methanol, although there is extensive evidence that several C_1 compounds can be oxidized by non-methylotrophic yeasts.

The ability to grow on methanol is not widely distributed in yeasts. In a survey in 1972, only 14 yeast species out of 422 tested in the Centraalbureau voor Schimmelcultures in Delft were found to grow on methanol. The ability seems to be confined to a few members of the genera *Candida*, *Hansenula*, *Pichia* and *Torulopsis*. The ability of yeasts to grow on methanol was first described by Japanese workers in 1969, followed rapidly by reports from Dutch and German laboratories. The various habitats from which methanol-utilizing yeasts have been isolated by enrichment culture showed that the best sources were flowers, soil and wood, which are all environments rich in methoxy groups from lignin and lignin decomposition products. *Pseudomonas putida* will degrade trimethoxybenzoate (a model lignin monomer) by converting the 3-methoxy group into free methanol which it does not oxidize, and which could thus be metabolized by other organisms. In this context it is noteworthy that many fungi that cannot grow on methanol can nevertheless oxidize it because they contain alcohol oxidase (see below).

Dissimilatory Pathway of Methanol in Yeasts

Methanol Oxidase. The first enzyme in the oxidation of methanol which converts it into formaldehyde is not, as in bacteria, a dehydrogenase, but an oxidase, of a nature that differs surprisingly little in the various species of yeasts from which it has been purified—3 strains of *Candida* and *Hansenula polymorpha*. Antibody to the enzyme from *C. boidinii* cross-reacts with enzymes from other species of yeast, but not with the enzymes from wood-rotting fungi. The enzyme, *alcohol oxidase* (EC 1.1.3.13) (Equation 4.1) is a flavoprotein containing non-covalently bound flavin-adenine dinucleotide (FAD). It consists of eight identical subunits, each of molecular weight about 80000, and eight moles of FAD per mole of enzyme. The enzyme is not specific for methanol. It will oxidize several alkanols and alkenols up to a carbon chain-length of C_4, but with decreasing affinity. Methanol with an apparent K_m value in the range 0.2 to 2 mM is the best substrate. The enzyme has fairly complex kinetics, indicative of a sequential mode of binding in which both substrates must form a complex with the enzyme. The apparent K_m for oxygen is rather high, in the region of 0.25 to 0.4 mM, which is higher than the normal concentration of oxygen in air-saturated buffer (0.2 mM). This means that the enzyme is always functioning under non-saturating substrate concentrations. This relatively low affinity for oxygen probably accounts for the high concentration of the enzyme in the yeast cell (see Fig. 4.2) which in many cases is 10% of the soluble protein of crude extracts of methanol-grown yeast cells, and possibly increases to well above 20% in certain circumstances. The enzyme also oxidizes formaldehyde. A mutant of *C. boidinii* lacking alcohol oxidase grows readily on glucose or ethanol, but will not grow on methanol. Yeasts grown on other substrates must synthesize the enzyme before growth on methanol can begin. The product of the reaction (Equation 4.1) is hydrogen peroxide

$$CH_3OH + O_2 \longrightarrow HCHO + H_2O_2 \tag{4.1}$$

and since this is a toxic metabolite it must be removed by catalase (Equation 4.2):

$$2H_2O_2 \longrightarrow 2H_2O + O_2 \tag{4.2}$$

Catalase (EC 1.11.1.6) activity is also elevated during growth on methanol compared with growth on ethanol. Catalase can also function peroxidatively to oxidize methanol, formaldehyde and formate in the presence of peroxide (Equation 4.3)

$$CH_3OH + H_2O_2 \longrightarrow HCHO + 2H_2O \tag{4.3}$$

The possibility that this is an additional route for methanol oxidation in yeasts cannot be entirely ruled out.

Formaldehyde Oxidation. The immediate product of methanol oxidation is formaldehyde, and this can be oxidized by methanol oxidase itself (probably because formaldehyde in solution consists mainly of methylene glycol $CH_2(OH)_2$). However it is generally thought that the major vehicle for formaldehyde oxidation in yeasts is an NAD^+- and glutahione(γ-glutamylcysteinylglycine, GSH) -dependent formaldehyde dehydrogenase (EC 1.2.1.1). The substrate of the enzyme is probably *S*-hydroxy-methylglutathione and in the reaction (Equation 4.4), the product is *S*-formylglutathione.

$$HCHO + GSH + NAD^+ \longrightarrow HCO\text{-}SG + NADH + H^+ \tag{4.4}$$

In *C. boidinii* this thiol ester is hydrolysed by a specific esterase (EC 3.1.2.12) (Equation 4.5) which has been purified. It is also present

$$HCO\text{-}SG + H_2O \longrightarrow GSH + HCOOH \tag{4.5}$$

in other yeasts. In *H. polymorpha* it has been suggested that *S*-formylglutathione is directly oxidized by the next enzyme in the sequence, formate dehydrogenase. Catalase can also oxidize formate (see above).

58

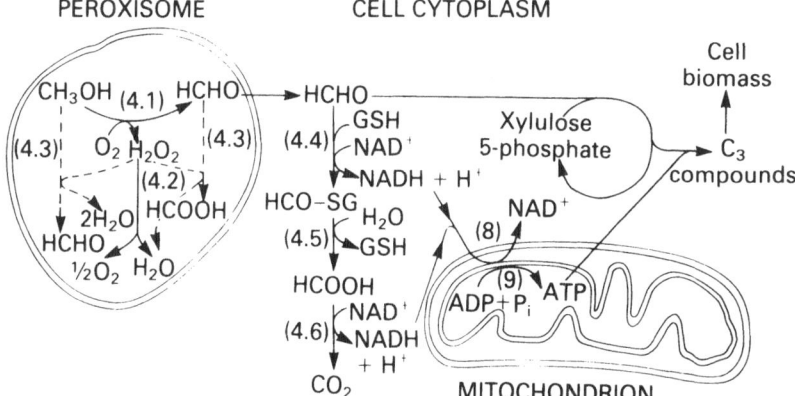

Fig. 4.1 Pathways of methanol oxidation, energy generation and carbon assimilation in yeasts in relation to subcellular structure. The numbers denote equations in the text: (4.1), methanol oxidase; (4.2), catalase; (4.3), catalase; (4.4), formaldehyde dehydrogenase; (4.5), S-formylglutathione hydrolase; (4.6), formate dehydrogenase; (4.7), dihydroxyacetone synthase; (8), NADH dehydrogenase on outer surface of inner mitochondrial membrane; (9), F_1 ATPase. (Based on J.P. Van Dijken, W. Harder and J.R. Quayle (1981) in H. Dalton (Ed.) *Microbial Growth on C_1 Compounds*, Proceedings of the 3rd International Symposium, 1981, by permission of John Wiley and Sons, Ltd.)

Formaldehyde dehydrogenase comprises 1% of the soluble protein in methanol-grown *C. boidinii.*

Formate Oxidation. Formate is oxidized to carbon dioxide in yeasts, as in bacteria, by an NAD^+-linked dehydrogenase (EC 1.2.1.2) (Fig. 3.9, reaction 6), but in yeasts the K_m for formate is extremely high (6 to 55 mM), whereas the K_m for S-formylglutathione is much lower. It has been suggested that in some yeasts the latter may be hydrolysed by the dehydrogenase with its immediate oxidation to CO_2, so that the overall reaction is as described in Equation 4.6

$$HCO\text{-}SG + NAD^+ + H_2O \longrightarrow CO_2 + GSH + NADH + H^+ \qquad (4.6)$$

Formate dehydrogenase comprises 3 to 5% of the soluble protein of methanol-grown yeasts. The overall scheme for the complete oxidation of methanol to carbon dioxide is thus as shown in Fig. 4.1.

Intracellular Location of the Enzymes of Methanol Oxidation

In most eukaryotic cells, catalase is not located free in the cytosol, but in defined organelles called *peroxisomes* or *microbodies*. In addition to catalase, many other enzymes, particularly oxidases that give rise to hydrogen peroxide as reaction product, such as glycollate oxidase, urate oxidase and glucose oxidase, are located in the peroxisomes. These organelles have important functions in eukaryotes, for example in the breakdown of glycollate, uric acid and fatty acids. Alcohol oxidase is also a peroxisomal enzyme. Cells grown on methanol contain many more (and larger) peroxisomes than

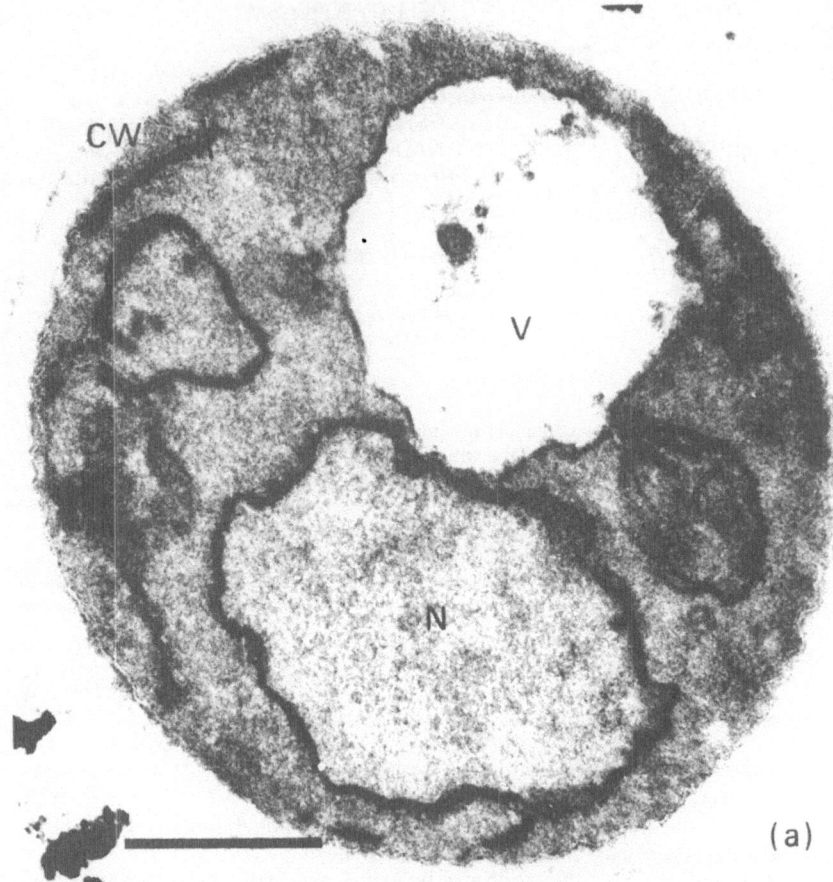

Fig. 4.2 Electron micrograph of thin sections of *Hansenula polymorpha*, (a) grown on glucose as sole carbon source; (b) grown on methanol. The bar represents 0.5 μm in each photograph. Note the great number of very large peroxisomes (marked P) in (b), in comparison with the single very small peroxisome in (a). Other symbols: CW, cell-wall; ER, endoplasmic reticulum; M, mitochondrion; N, nucleus; V, vacuole. (Photographs kindly provided by Dr. M. Veenhuis, Laboratory of Electron Microscopy, Biological Centre, University of Groningen, The Netherlands.)

cells grown on glucose (Fig. 4.2), and specific staining reactions have been used to demonstrate conclusively that methanol oxidase is present in the peroxisomes. Moreover it has been shown that the crystalline structure visible in peroxisomes at high magnification with the electron microscope is actually due to the presence of crystals of alcohol oxidase. Thus peroxisomes play a special role in the methylotrophic growth of yeasts on methanol. Formaldehyde, the presumed product of peroxisomal methanol oxidation, must pass into the cytosol before it can be oxidized further (Fig. 4.1). All the

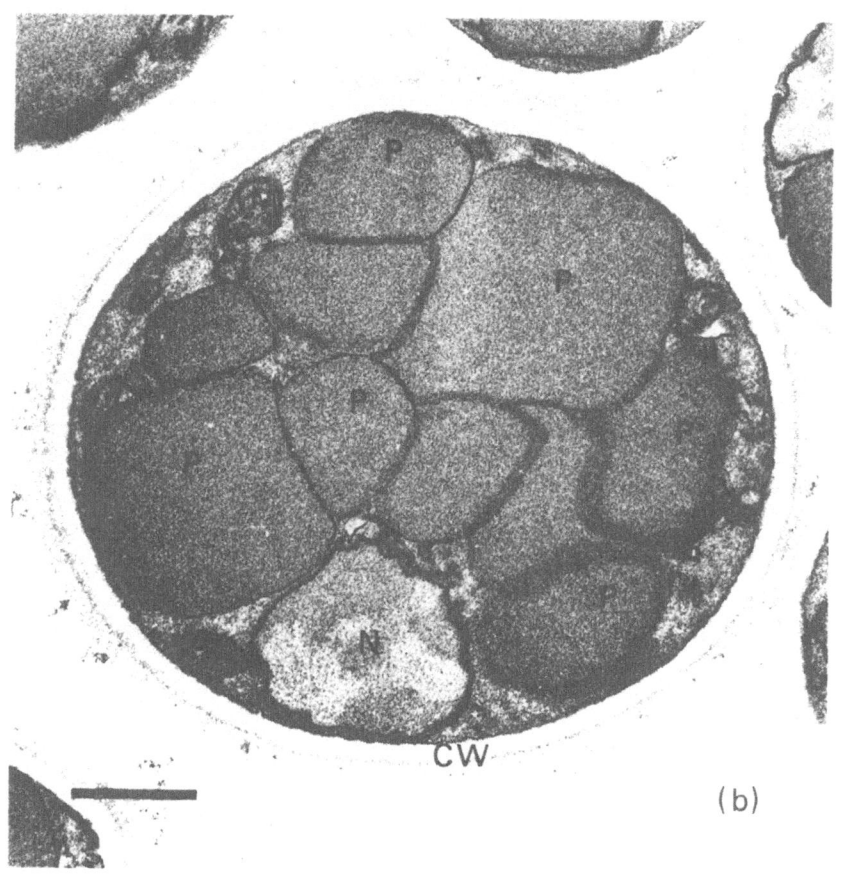

(b)

NADH required for the generation of ATP (see Chapter 3) is thus formed outside the mitochondria. This NADH is oxidized by an NADH dehydrogenase located on the outer surface of the inner mitochondrial membrane and gives rise to ATP in exactly the same way as NADH generated from other substrates by oxidation within the mitochondrion. The only difference is that the outward-facing NADH dehydrogenase seems not to involve the first of the sites (or loops) of oxidative phosphorylation with the result that one molecule of NADH generates two rather than three molecules of ATP.

Assimilatory Pathway of Methanol in Yeasts

In yeasts, as in bacteria, formaldehyde, the first product of methanol oxidation, is the precursor of cell material. Recent work has concentrated on characterizing the pathway by which formaldehyde enters cell material. Experiments with *Candida* N-16 and *H. polymorpha* showed that the first labelled products when [^{14}C]methanol was given to yeasts were sugar phosphates, and the same initial labelling pattern was also obtained with [^{14}C]formaldehyde. This pattern resembled very closely the results seen when bacteria containing the hexulose phosphate pathway were given labelled methane or methanol (Chapter 3). Thus for several years it was believed that yeasts had the same hexulose phosphate carbon assimilation pathway (Fig. 3.4) as bacteria. This idea was weakened when the key enzymes of the pathway were sought. Hexulose phosphate isomerase (Equation 3.5) could not be detected in methanol-grown yeasts and the activity of the other key enzyme of the cycle, hexulose phosphate synthase (Equation 3.4) was very low indeed. Recent work by Quayle and his colleagues and by Japanese workers suggests that *in vivo* a totally different biosynthetic pathway for methanol incorporation operates in yeasts, a pathway called the *dihydroxyacetone pathway*, since this is the first labelled product. The key reactions of this pathway are as follows (Fig. 4.3). The initial formaldehyde acceptor is xylulose 5-phosphate, and the first enzyme in the pathway is a novel type of transketolase called *dihydroxyacetone synthase* (Equation 4.7)

HCHO + Xylulose 5-phosphate $\longrightarrow$
$\qquad$ Dihydroxyacetone + Glyceraldehyde 3-phosphate $\quad$ (4.7)

Dihydroxyacetone + ATP $\longrightarrow$ Dihydroxyacetone phosphate + ADP $\quad$ (4.8)

Dihydroxyacetone phosphate + Glyceraldehyde 3-phosphate $\longrightarrow$
$\qquad$ Fructose 1,6-bisphosphate $\quad$ (4.9)

Fructose 1,6-bisphosphate + H$_2$O $\longrightarrow$ Fructose 6-phosphate + P$_i$ $\quad$ (4.10)

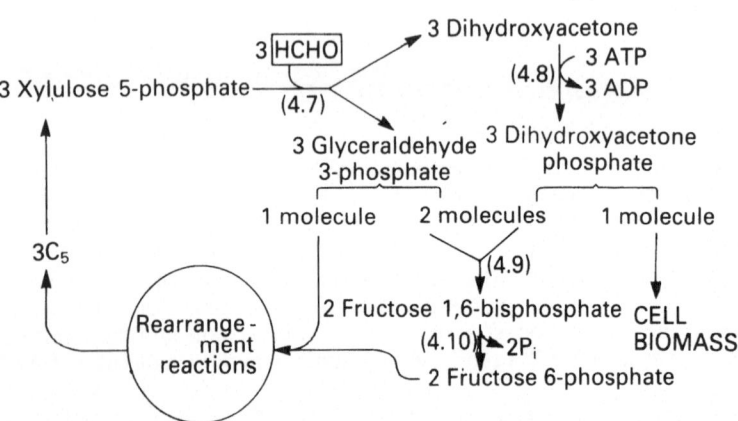

Fig. 4.3 The dihydroxyacetone pathway of formaldehyde fixation in methylotrophic yeasts. The numbers in brackets denote the equations in the text. The enzymes are (4.7), dihydroxyacetone synthase (a specialized transketolase); (4.8), triokinase; (4.9), fructose bisphosphate aldolase; (4.10), fructose bisphosphatase.

This primary fixation route is characterized by two novel enzymes, dihydroxyacetone synthase and *triokinase* (Equation 4.8) (EC 2.7.1.30). These steps are then followed by a cycle of rearrangement reactions which regenerate the xylulose 5-phosphate from fructose 6-phosphate, and are basically similar to those in Fig. 3.5.

The overall result of this sequence is thus the formation of a molecule of dihydroxyacetone phosphate from three molecules of formaldehyde and three of ATP (Equation 4.11).

$$3HCHO + 3ATP \longrightarrow \text{Dihydroxyacetone phosphate} + 3ADP + 2P_i \quad (4.11)$$

Careful experiments have demonstrated the formation of [14C]dihydroxyacetone in crude extracts of *C. boidinii* given xylulose 5-phosphate and [14C]formaldehyde, and it has been shown that the activities of all the enzymes catalysing the reactions of Equations 4.7–4.10 are elevated in cells grown on methanol compared with those grown on glucose.

Possible Applications of Methylotrophic Yeasts

The possibility of using yeast as a source of single-cell protein (see Table 5.2) has been considered. Although yeasts have several advantages over bacteria for this purpose — they have a lower nucleic acid content, and recovery of the larger cells from the growth medium is easier — they have some serious disadvantages. The protein content of bacteria is 70% of the dry weight compared with only 45% in yeasts, and yields of yeast cells are also much lower compared with bacteria (yeasts 0.35 to 0.4 g dry cells per g methanol compared with 0.55 g for certain bacteria).

Alcohol oxidase may prove to have useful applications in the analytical field. An electrode incorporating immobilized alcohol oxidase as a means of measuring alcohol has recently been described.

Summary

The pathways of assimilation and dissimilation of methanol by methylotrophic yeasts (species of *Candida*, *Pichia*, *Hansenula* and *Torulopsis*) are different from the corresponding pathways in bacteria. Methanol is oxidized in the peroxisomes by a flavoprotein alcohol oxidase. The formaldehyde formed enters the cytoplasm where it can be further oxidized or converted to cell material. ATP formation depends on an NADH dehydrogenase located on the outside of the inner mitochondrial membrane. The assimilation pathway is characterized by the reaction of formaldehyde with xylulose 5-phosphate to give glyceraldehyde 3-phosphate and dihydroxyacetone. The latter must be phosphorylated by triokinase before it can act as a precursor of sugar phosphates, which are the first radio-active compounds to be detected when [14C]methanol is administered to whole cells. The dihydroxyacetone pathway of methanol assimilation is an important new biosynthetic route in these yeasts. Methylotrophic yeasts offer good opportunities for studying the role of peroxisomes in the physiology of the cell.

References

COONEY, C.L. (1975). Engineering considerations in the production of single-cell protein from methanol. In *Microbial Growth on C_1-Compounds*. Proceedings of the International Symposium, Tokyo. 1974. pp. 183–197. Society of Fermentation Technology, Osaka, Japan.

FUJII, T., ASADA, Y. and TONOMURA, K. (1975). Assimilation of methanol by *Candida* Species N-16. In *Microbial Growth on C_1-Compounds* (as above). pp. 121–137.

SAHM, H. (1977). Metabolism of methanol by yeasts. *Advances in Biochemical Engineering* 6, 77–103.

TANI, Y., KATO, N. and YAMADA, H. (1978). Utilization of methanol by yeast. *Advances in Applied Microbiology* 24, 165–186.

VAN DIJKEN, J.P., HARDER, W., BEARDSMORE, A.J. and QUAYLE, J.R. (1978). Dihydroxyaceton: an intermediate in the assimilation of methanol by yeasts? *FEMS Microbiology Letters* 4, 97–102.

VAN DIJKEN, J.P., HARDER, W. and QUAYLE, J.R. (1981). Energy transduction and carbon assimilation in methylotrophic yeasts. In *Microbial Growth on C_1 Compounds*. Proceedings of the 3rd International Symposium, Sheffield, 1980. pp. 191–201. Edited by H. Dalton. Heyden & Son, London.

WAITES, M.J. and QUAYLE, J.R. (1980). Dihydroxyacetone: a product of xylulose 5-phosphate-dependent fixation of formaldehyde by methanol-grown *Candida boidinii*. *Journal of General Microbiology* 118, 321–327.

5 Biotechnological Applications of Methanogenic and Methylotrophic Micro-organisms

In the early 1960s, the physiology and biochemistry of micro-organisms capable of growth on C_1 compounds was an obscure field of biology, of interest only to those who wanted to examine the borderline between autotrophy and heterotrophy. At that time also only a few species of methanogenic bacteria had been obtained in pure culture, and our knowledge of the biochemistry of methanogenesis was very slight. One of the reasons for the subsequent tremendous development in our knowledge of the two areas described in Chapters 2 to 4 is the realization of the wide range of possible technological applications of methanogenic and methylotrophic organisms. This chapter covers the more important of these applications, which range from quite old techniques (such as the anaerobic sludge digester) to applications that although potentially of great usefulness have still not left the laboratory stage of development. The topics treated in the greatest detail, because of their established commercial and social usefulness, are the production of methane from organic waste and the use of methylotrophs for the production of animal feedstuffs.

The Production of Methane from Organic Waste

If organic materials are held under anaerobic conditions a bacterial population develops that is similar in many ways to that in the rumen or in anaerobic fresh-water mud (Chapter 2). Such a population can break down organic waste materials and convert them into a gas containing 2 volumes of methane per volume of CO_2, together with a solid residue that is virtually odourless and rich in nitrogen, phosphorus and metal ions. The gas has been given various names — sewer gas, sludge gas, marsh gas, etc. — but it has recently become fashionable to call it *biogas*. Despite its content of CO_2, it will burn readily, but requires a high temperature to ignite it (700 °C) whereas for methane the ignition temperature is 645 °C. The solid residue contains 25% protein-nitrogen (from the bacterial cells), about 3% ammonia-nitrogen, 9% fat, 6% phosphate expressed as P_2O_5 and 1.4% potassium expressed as K_2O. It thus makes a good fertilizer.

Many applications of anaerobic 'digestion' of organic material are now in use or are being considered (Table 5.1), and these have different aims, although it is often possible to combine several of these. The kind of installation used varies considerably from enormous plant to process sewage from cities of several million people, to small simple units for use either in the developed countries to dispose of evil-smelling faecal residues from intensive pig, cattle and poultry units with the production of both odourless fertilizer and gas that can be used on the farm, or for use in developing countries like China and India. Here such installations may be used to remove large

Table 5.1 Applications of the Anaerobic Digestion of Organic Materials

Sewage purification, including removal of faecal pathogens.
Concentration of sewage sludge before an oxidative biological treatment process (activated sludge).
Disposal of offensive odorous excrement from intensive pig and poultry units.
Production of useful combustible biogas.
Production of a rich odourless fertilizer.
Disposal of large quantities of cattle excrement.
Disposal of crop residues (straw, palm-oil waste, etc.).
Disposal of waste from food-processing plants (dairy, brewing, potato, sugar-beet, etc.).
Disposal of waste from the wool-textile industry.
Disposal of domestic refuse (the organic part).
Conversion of agricultural crops to biogas fuel.

quantities of human and cattle faecal residues with their conversion to odourless fertilizer with the biogas being either used or allowed to escape depending on the needs of the society. By such means the control of pollution and epidemics can be combined with the production of fertilizer and energy.

The basic design of the digester is similar in principle in all applications, although there will be variations in the type of gas-collecting reservoir and the mode of operation, which may be continuous, semi-continuous or batch. The digester consists of a concrete tank with entry and overflow pipes, preferably wholly or partially underground. There is a domed roof of concrete or cast-iron (or flexible and moveable) over the gas-collecting space, with an exit valve for the gas. The operating temperature (i.e. whether a mesophilic or a thermophilic anaerobic population is used) will also vary with the location of the installation. Despite optimistic predictions, the biogas yield is in most cases not sufficient to make its purification and transport worthwhile. Biogas tends to be used almost exclusively on the site where it is formed, either for domestic heating, heating for the fermenter (especially at sub-zero temperatures) or to drive generators to make electricity.

Sewage Purification. This is the oldest application of anaerobic digestion. The first anaerobic digester was built in 1895 for the city of Exeter, England by Donald Cameron. The methane produced was used for lighting in the area of the sewage plant. In Britain anaerobic digestion is used to treat about half the sewage produced, this mainly in the large conurbations. In the USA the process is widely used in the major cities.

After the sewage enters the treatment works (Fig. 5.1), it moves slowly through primary sedimentation tanks to separate the sewage sludge from the supernatant. The supernatant then passes through further processing involving aerobic bacteria ('activated sludge') before being clean enough to discharge into a river. By 'clean' is meant having a low content of organic materials and ammonia so that the *biological oxygen demand*, i.e. oxygen consumed when the organic material is metabolized or the ammonia converted into nitrate by micro-organisms, is relatively low (see Chapter 1). If an anaerobic digestion process is used, the sludge from the primary

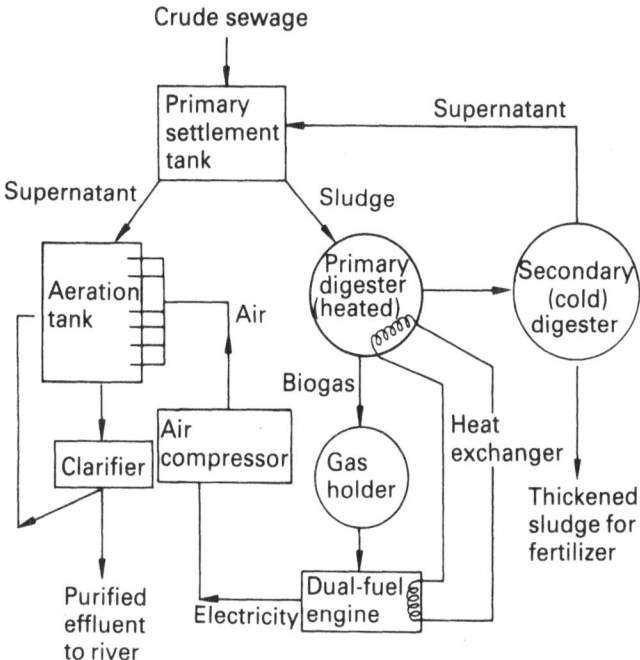

Crude sewage

Primary settlement tank

Supernatant

Supernatant

Sludge

Aeration tank

Air

Primary digester (heated)

Secondary (cold) digester

Biogas

Air compressor

Clarifier

Gas holder

Heat exchanger

Thickened sludge for fertilizer

Purified effluent to river

Electricity

Dual-fuel engine

Fig. 5.1 The role of anaerobic digestion in sewage purification (adapted from F.E. Mosey (1980) in D.A. Stafford, B.I. Wheatley and D.E. Hughes (Eds.) *Anaerobic Digestion*, by permission of Applied Science Publishers).

sedimentation is pumped to a digester where it has a residence time of 20–30 days, although shorter periods of about 10 days are equally effective. The digester is a reinforced concrete tank of capacity 10^3 to 10^4 m^3 which should be well insulated. The tank is mixed to ensure a turnover every 1 to 4 hours and is warmed, usually to a temperature of about 35 °C.

The biogas obtained is not usually transported, but used on site either as fuel for boilers for warming the digester, or is burned in dual-fuel engines in which diesel oil provides 5 to 10% of the engine power, the remainder coming from the biogas. This is fed in mixed with air with a gas:air ratio of 1:7. The engines may be coupled directly to compressors to provide the compressed air needed for the activated sludge (aerobic) part of the process or used to drive generators to produce electricity, which can be used for all the machinery in the sewage works, and if in excess can be sold to the national electricity network. The dual-fuel engines are only about 35% efficient, but the energy released as heat is not wasted, but used to warm the digester tank (Fig. 5.1). For a 25-day retention period in a digester with a 9% (w/v) suspension of solids, 35% of the solids are destroyed and 80% of the grease, with the formation of about 1.1 m^3 of biogas per kg of solids destroyed. The sludge from the digester is a mixture of undegraded material enriched by bacterial biomass, and its best use is as a fertilizer. It is still unsuitable for discharge into rivers, but is not unpleasant to handle.

67

Manure Disposal on Farms. Animal excrement is a wet, heavy, unpleasant smelling, but inevitable, product of animal husbandry. With modern intensive livestock rearing, the animals no longer spread it conveniently over the fields; it accumulates in the animal houses. In cases of pig or poultry excrement, it is particularly foul smelling. Transport costs prevent the removal of this material, which must thus be used on the farm. If such material is mixed with water to give a slurry it can be sprayed directly on to the fields as fertilizer, where it often gives rise to offensive odours and may pollute streams and waterways. To minimize pollution of this sort, large farms are being encouraged to use anaerobic digestion to turn excrement into odourless fertilizer with the simultaneous production of biogas. The gas can be used for domestic heating, agricultural heating and electricity generation, as well as for warming the digester vessel. The fertilizer produced is as rich in nitrogen, phosphorus and potassium as the original excrement, is easier to handle, relatively odourless and has better soil-conditioning properties. Anaerobic digesters must be well insulated, particularly if designed for use on American farms where temperatures may fall to $-30\,°C$ in the winter. The economics of such systems are well established for large farms (250 fattening pigs or 500 head of cattle). For the latter, estimates suggest that the use of biogas as a fuel can produce savings sufficient to pay for the installation in two years (Hayes et al., 1980). In the future increasing fuel costs may make biogas production also worthwhile on smaller farms. This is in addition to the role of the digester in pollution prevention and fertilizer production. The functioning of the digester is such that it will often tolerate large quantities of other organic material such as straw, though probably not wood-shavings.

Digesters also have applications in the developing nations. They have been used with great success in China and with rather less success in India. The successful Chinese installations have been on a domestic rather than a community scale. Digesters are used in the third world for the production of fertilizer, the prevention of epidemics by the treatment of excrement and the disposal of certain agricultural wastes rather than the production of biogas, which many poor communities lack the resources to use. In any case we need to remember that cattle dung and rice-straw are not useless products in the developing countries, but valuable fuel resources in themselves. It is perhaps the public-health aspects of anaerobic digestion that are the most valuable.

In all applications in which biogas is generated, there is a limited explosion hazard. This is not usually dangerous because the high ignition temperature of biogas means that relatively large amounts of air must be present. The presence of air at up to 25% by volume is quite harmless. What is important with biogas plants is to install good ventilation so that leaks may escape harmlessly, and to site gas storage reservoirs in safe positions. Sewage biogas-handling equipment always contains flame traps.

Crop-Waste Disposal. In some countries the anaerobic treatment of wastes from crop- and food-processing plants (potatoes, sugar beet, etc.) is already in practice. Such processes will become increasingly important everywhere with the increasingly rigorous restrictions on discharge of such wastes into waterways. The industries involved are likely to be attracted to a process that in addition to preventing pollution may allow the production of saleable fertilizer and biogas.

Conversion of Crops to Biogas. Some countries that rely for fuels on imported oil are experimenting with the replacement of fossil fuels by fuels made from agricultural crops, which are a renewable resource. Thus Brazil is developing processes to turn sugar-cane and manioc plants via fermentation and distillation into industrial ethanol for use as an automobile fuel. New Zealand is considering growing crops for anaerobic conversion to biogas. The sludge also formed could be used as a fertilizer to help to restore the nutrients depleted by the growing crops. Such processes may be regarded by some as a waste of agricultural resources, but the process may yet become economically worthwhile. Whole crops, including parts such as straw, which might otherwise be wasted, can be used, and the crops can be used green. Silage can be used as the input for the digester and facilitates digestion because volatile fatty acid formation begins in the silage stock.

Anaerobic Digestion in the Disposal of Domestic Refuse. Apart from the recovery of metals, most new processes for the disposal of rubbish concentrate on turning it into energy, since energy costs are continually increasing. Domestic refuse is continually increasing in quantity, and disposal processes continue to rise in cost. Modern methods of refuse management all concentrate on trapping some of the potential energy of refuse into a useable form. Numerous combustion and pyrolysis techniques have been tried, and some combustion plants are associated with district-heating or energy-generation installations. The conversion of organic refuse into biogas is another possible approach.

Pilot plant operations of this sort are being experimented with in the USA. After removal of the 27% of domestic refuse that is glass, plastic and metals, the remaining material, rich in organic compounds, is shredded and slurried with water and run into an anaerobic digester. Biogas is formed. The final step is the separation, dewatering and disposal of the fermented sludge by landfill or incineration. The biogas can be used for processes on the site. This approach turns refuse disposal into a similar process to sewage disposal. The processing costs are high, however, and without the public-health pressures that led to proper methods of sewage disposal, the production of biogas from refuse is probably a non-economic process.

Single-Cell Protein

The increasing size of the world's population creates a demand for protein that agriculture is increasingly unable to satisfy despite increases in productive efficiency. This shortfall between supply of protein and demand has been called the *protein gap*. This problem presents two different aspects. In the developed countries, increasing standards of living create an increased demand for more foods rich in protein, such as meat, eggs and dairy products, and in developing countries the increase in population creates an increased requirement for protein at a more basic nutritional level. Often these two factors come into conflict; for increased transport efficiency means that protein-rich foods, such as soyabean meal produced in the developing lands and fish caught there, instead of being used to provide protein for the expanding indigenous populations, are exported to the developed countries where they are not directly used as human food, but as feedstuffs for poultry and domestic animals.

Intensive modern animal husbandry for the production of eggs, poultry, veal and pork requires strictly formulated feedstocks. These compounded animal feeds contain 10–30% protein, depending on their use, and protein content is the prime factor determining feed quality and performance. Traditionally, soyabean meal and fish meal have been the main components of these feedstocks. The quality of a feedstuff is determined by the *biological value* if the protein in it: for optimal nutrition of non-ruminant vertebrates the protein must have an adequate content of essential amino acids (which cannot be synthesized by vertebrates), particularly lysine, methionine, threonine and tryptophan. Soyabean meal (45% protein by weight) and fish meal (65% protein by weight) are both of high biological value. Evidence suggests that little further growth can be expected in the world fish catch, and more of the fish that is caught is being used directly for human food. Similarly, although world production of soyabean meal doubled between 1963 and 1973, there is now evidence that unless there is a dramatic increase in the arable acreage devoted to growing soyabeans, or a massive increase in yields as a result of the use of artificial fertilizers, world production may not increase so rapidly. In any case, the arable acreage devoted to the 60% of the world production of soyabeans that takes place in the USA could at any time be switched to other crops if these proved to be more profitable.

Microbiologists and technologists over the last 20 years have been developing alternative protein sources using micro-organisms. In the developed lands these could serve as a substitute for soyabean meal and fish meal, thus making more of these materials available to the developing nations. Alternatively, processing methods may be developed that, by reducing the nucleic acid content, will make microbial protein directly acceptable for human food, although this has not yet occurred to any extent. This development will be helped if microbial protein is found to possess *functional applications*, i.e. capacity to be whipped or form gels, etc., that would make it useful for specialized food applications. Protein from microbial sources is called *single-cell protein.*

Carbon Sources for Single-Cell Protein Production. A microbial source of carbon for single-cell protein production should ideally not be based on a material, such as starch, that is itself a valuable food source, unless circumstances require the 'up-grading' of carbohydrate into protein. The ideal objective is to create a new food chain by growing micro-organisms on carbon sources that otherwise have no nutritional value. The ideal raw material would be something with no commercial value, such as agricultural wastes (e.g. straw) or wastes from food-processing industries that (as we saw above) present a pollution hazard in their disposal. Other materials in this category are sewage, animal excrement, wood-shavings, etc. Although these products may often require expensive pre-processing before they can be used as growth substrates for micro-organisms, several single-cell protein production processes have been attempted on a laboratory scale. There may be problems associated with separation of the product from the growth substrate, and in any case this topic is outside the scope of this book.

The next best choice of carbon source would be rather cheap organic compounds, readily available and either without any other industrial application or with very limited alternative uses. In the sixties some petroleum products (long-chain hydrocarbons), natural gas and some petrochemicals came into the category of cheap raw materials. Political and economic changes in the world since then (particularly the spectacular rise in the price of oil) have meant that these are no longer very cheap. Nevertheless, increases in the price of soyabean meal and fish meal have meant that as a potential feedstuff for animal husbandry, single-cell protein may still compete effectively in price. Moreover the price of such a product is not subject to the irrational political oscillations of the world commodity markets, and could help to reduce import costs. For example in 1973 the EEC nations spent £1600 million on importing 5.7 million tonnes of high-quality protein meal.

When selecting a growth substrate for single-cell protein, the overriding economic consideration is the *yield coefficient*. This is the mass of dry microbial cells (*biomass*) produced per unit mass of substrate (carbon source) utilized. It depends on both the chemical composition of the carbon source and the efficiency of utilization of the substrate by the microbial culture employed. The latter depends on environmental conditions in the fermenter and on the mode of operation. It also depends on the type of micro-organism used and its pathway of carbon assimilation (see below). In order to compare different growth substrates, the yield coefficient is best expressed as the *carbon yield coefficient*, i.e. the mass of dry cell material produced per unit mass of substrate carbon used. For long-chain alkanes (C_{12} to C_{18}), methane, methanol and ethanol these figures are, respectively, approximately 1.19, 1.06, 1.33 and 1.34 g biomass/g carbon. This figure also depends on growth conditions. If we assume that the microbial cells have a similar elemental composition irrespective of growth substrate (see below), the figures suggest that alcohols are more efficiently utilized for microbial growth than hydrocarbons. The use of long-chain hydrocarbons and ethanol is outside the scope of this book, but the use of hydrocarbons for single-cell protein production was pioneered by BP. Their product, Toprina, was manufactured for 10 years. Political developments in Japan and Italy have retarded the large-scale use of Toprina (see Laskin, 1977). It was a product of the yeast *Candida lipolytica*.

Methane and methanol, which are industrially by far the most important reduced C_1 compounds, have both been considered as substrates for the production of single-cell protein. Methane has the advantage that it is cheap, and despite its employment as an energy source, large amounts are still wasted in various parts of the world because it is uneconomic to collect it. The few impurities it contains are gaseous in nature, and there is no residue left behind in the growth medium. However, methane–air mixtures can be explosive, and bacteria growing on methane grow less satisfactorily in pure culture than they do, for example, on methanol. Although Shell has accumulated extensive knowledge and developed some novel techniques for obtaining single-cell protein from methane, viable commercial processes have not yet been developed.

Synthesis-, water- and blast furnace-gas are all mixtures of carbon monoxide and hydrogen. Synthesis-gas is an intermediate in some processes of methanol production. These gases could also, in the presence of oxygen or air, be used for single-cell protein production. Production of 1 tonne of pig-iron results in the formation of 7 tonnes of blast furnace-gas, which is essentially a waste product. Pilot experiments have shown that the carbon monoxide-utilizing bacterium *Pseudomonas carboxydovorans* (Chapter 3) will grow well on a mixture of synthesis gas and air (50:50 v/v). If the cells are pre-adapted to growth on H_2, they grow autotrophically by oxidation of H_2, whereas if they are pre-adapted to growth on CO, they grow by obtaining energy by oxidation of the CO. Theoretically 300 tonnes of dry cell mass per day could be produced from the waste gases of each blast furnace. The practicalities of these possibilities on an industrial scale remain to be investigated (see Meyer, 1980).

Single-Cell Protein from Methanol. The result of a decade of experimentation has been that at least four different companies have started production, on various scales, of single-cell protein, mainly from bacteria grown on methanol as sole source of carbon. Methanol offers several advantages over alternative sources of carbon:

(a) it is very soluble in water, which avoids the phase-transfer problem associated with hydrocarbons;
(b) unlike methane–air mixtures, there is little risk of explosion;
(c) it is readily available, commercial processes exist to synthesize it from various hydrocarbon sources such as methane, naphtha, synthesis gas and coal;
(d) it is readily purified, and some methods of preparation avoid the risk of contamination by toxic polycyclic aromatic compounds;
(e) it requires less oxygen for its metabolism by micro-organisms than does methane and there is a correspondingly smaller need for cooling during growth.

Bacteria, yeasts and a few strains of fungi are known which can grow on methanol (Chapters 3 and 4). The choice of organism for single-cell protein production thus lies between bacteria and yeasts. The advantages and disadvantages of these are summarized in Table 5.2. Their higher biomass yields, faster growth rate and higher protein content have made bacteria in most cases the organisms of choice (at least as far as animal feedstuffs are concerned). There are several theoretical reasons why this should be so.

Growth Yield and the Selection of Micro-Organisms for Single-Cell Protein Production from Methanol. In single-cell protein work, the carbon source for the fermentation is often the most expensive item in the running costs (see below). This means that maximization of the cell yield is of great importance, and has been largely responsible for the revolutionary designs of fermenter vessels (see below) to improve growth conditions. Extensive work over the last 25 years has enabled us to arrive at theoretical

Table 5.2 Advantages and Disadvantages of Yeasts and Bacteria for the Production of Single-Cell Protein from Methanol

Adapted from E.L. Krug, H.C. Lim and G.T. Tsao (1979) *Annual Reports on Fermentation Processes* **3**, 141–195, by permission.

	Bacteria	Yeasts
Protein content (% of dry weight)	50–70	45–55
Doubling time (h)	1.4	3–10
Nucleic acid content (% of dry weight)	10–20	7–12
Size (μm)	1	5–10
Typical growth yield on methanol (g biomass/g methanol)	0.55 (hexulose phosph. pathway) 0.40 (serine pathway)	0.44
Growth	Simple medium	Vitamins needed
Contamination problems	Yes	Minimal
Amino acid distribution (g amino acid/100 g dry biomass)		
FAO reference protein*		
Methionine 2.2	2.0	0.41–0.8
Lysine 4.2	4.9	2.1–4.1
Ease of harvesting	Poor	Good
Acceptability for human consumption	Poor	Reasonable
Elemental composition (approx.)	$C_4H_8O_2N$	$C_4H_7O_2N_{0.6}$

*Nutritional protein standard of the United Nations Food and Agriculture Organisation.

growth yields for any particular carbon source whose pathway of metabolism in the cells is completely known. This field is rather complex, and the following discussion is only a simplified introduction: for detailed discussion of this topic see Harder and Van Dijken (1976), Anthony (1978), Stouthamer (1979) and Harder *et al.* (1981).

Experiments using substrates (mainly sugars) whose mode of catabolism and whose ATP yield (mol ATP formed metabolically/mol substrate catabolized) are known has enabled us to calculate Y_{ATP}, the number of grams of biomass (cell dry weight) formed per mol ATP generated during catabolism. From such experiments, an average value of Y_{ATP} has been arrived at which is about 10.5 g biomass/mol ATP. There are good experimental grounds for assuming that the synthesis of cell material from 3-phosphoglycerate requires approximately the same amount of ATP as synthesis from glucose. It is thus possible to arrive at an estimate of the energy requirement for the synthesis of cell constituents from 3-phosphoglycerate assuming a Y_{ATP} of 10.5. The reducing power required (i.e. NADH or NADPH) can also be worked out if the elemental composition of the cell biomass is known. If in addition we know what the energy requirements are to make 3-phosphoglycerate from reduced C_1 compounds (and this depends on the pathway involved, see

Table 3.2), then we should be able to predict the biomass yield (of a particular composition) per mole of C_1 growth substrate utilized from the energy produced by oxidation of the substrate. Most calculations are based on data that show that bacterial biomass contains elements in the proportions $C_4H_8O_2N$. The corresponding molar elemental proportions for yeasts are $C_4H_7O_2N_{0.6}$. The calculation requires that we know how many moles of ATP are formed per mole of substrate oxidized. Because of our, as yet, incomplete knowledge of bacterial electron-transport pathways, this information is not always available, but since the number of possible values is limited, it is possible to make informed guesses. The topic is considered by Drozd and Wren (1980). For yeast growing on methanol, the oxidation of methanol to formaldehyde produces no ATP (see Chapter 4) and the oxidation of formaldehyde to CO_2 via dehydrogenases would produce 2NADH, equivalent to 4 molecules of ATP (Chapter 4). Despite our uncertainties as to the pathways of methanol and formaldehyde oxidation in bacteria (see Chapter 3), the fact that methanol is oxidized via a dehydrogenase rather than an oxidase means that more ATP per mol methanol is almost certain to be produced. We would thus predict that bacterial growth yields with methanol as substrate would be higher than those for yeasts, and Table 5.2 shows that this is true experimentally for bacteria using the hexulose phosphate pathway. The higher energy requirements (Table 3.2) for serine pathway bacteria means that they have a lower growth yield (Table 5.2). All the bacteria selected for the production of single-cell protein from methanol (Table 5.3) possess the hexulose phosphate pathway. Other important criteria are lack of pathogenicity and the production of protein of high biological value (yeast protein tends to be deficient in methionine, Table 5.2).

The Pruteen Process. Of the four companies who have experimented with the production of single-cell protein from bacteria and yeasts grown on methanol (Table 5.3), only ICI has gone into commercial production. Current production targets in 1980–81 were 40 000 tonnes of Pruteen per year. The plant has a maximum production capacity of 50 000 tonnes per year. The product (in December 1980) sold at the same price as fish meal, and about double that of soyabean meal (the protein content of Pruteen is

Table 5.3 Experimental Single-Cell Protein Produced by Various Companies from Methanol-Grown Bacteria and Yeasts

Parent companies	Subsidiary	Organism used	Name of product
Imperial Chemical Industries (UK)	—	*Methylophilus methylotrophus*	Pruteen
Uhde/Hoechst (W. Germany)	—	*Methylomonas clara*	Probion
AB Marabou/Norsk Hydro	Norprotein (Sweden)	*Methylomonas methanolica*	Norprotein
Mitsubishi Gas Chemical Co. (Japan)	—	*Methylomonas methanolis BNK-84*	—
		Yeast	—

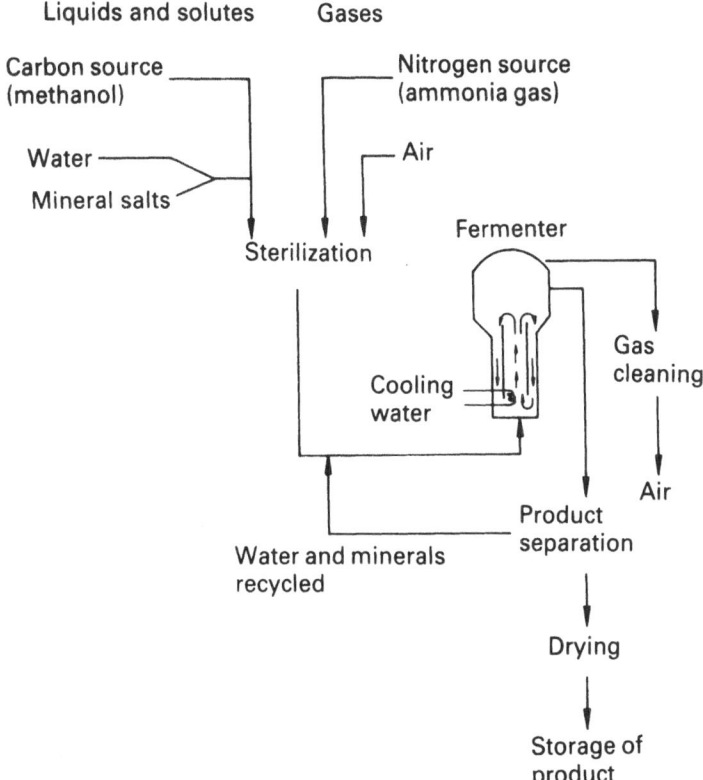

Liquids and solutes **Gases**

Carbon source
(methanol)

Nitrogen source
(ammonia gas)

Water —

Air

Mineral salts

Fermenter

Sterilization

Gas
cleaning

Cooling
water

Air

Product
separation

Water and minerals
recycled

Drying

Storage of
product

Fig. 5.2 Process flow diagram for single-cell protein production from methanol using the ICI pressure-cycle fermenter.

also nearly double that of soyabean meal). A flow diagram of the plant is given in Fig. 5.2. It resembles closely the process of large-scale growth of micro-organisms as carried out in a laboratory fermenter. The whole process is carried out completely aseptically so as to prevent contamination by foreign organisms. This is essential in order to ensure that the product remains identical with the organism on which extensive (and expensive) toxicity tests were carried out. There is thus none of the rather casual conditions common in many other fermentation industries employing, for example, open fermenters.

The amount of heat produced during growth of micro-organisms is proportional to the oxygen consumption, so hydrocarbon fermentations produce more heat than processes where alcohols are the growth substrate.

The high oxygen transfer rates and the necessity for efficient heat removal that are essential for growth of micro-organisms have resulted in a requirement for completely novel types of fermenter vessel, and single-cell protein processes have led to a series of innovative developments in plant design. The two most important of these are the air-lift fermenter and ICI's special development of it, the pressure-cycle fermenter. The latter consists of a closed, totally aseptic fermenter (with aseptic valves and entrance points) in two parts. The first is an ascending limb or 'riser' into which compressed air and sterile medium enter. The incoming gases (air and ammonia, which is the nitrogen source) are sterilized by filtration, the liquid medium by heat. At both top and bottom the riser is connected to a descending limb, the 'downcomer'. The riser contains a two-phase mixture of air and culture moving at a rapid velocity. Most of the oxygen transfer takes place here, and the aeration and mixing are so thorough that 50% of the incoming oxygen dissolves. The dry weight of the cell suspension in the fermenter is about $30\,g/l$, and growth takes place at an operating temperature of $40\,°C$. At the top of the fermenter the air becomes exhausted and the hydrostatic pressure decreases. The spent air leaves at the top, and the liquid flows into the downcomer. The heat-exchanger for cooling of the medium is situated at the base of the downcomer. At the bottom the cooled fluid is directed into the riser, picks up fresh air and completes the pressure cycle. The circulation is driven by the hydrostatic pressure difference between the broad diameter riser and the narrow diameter downcomer.

The continuous stream of culture leaving the fermenter is subjected to a temperature/acid shock that separates flocculated cells from the clear supernatant. This liquid is adjusted in composition and recycled to the fermenter. 99% of the biomass is recovered, dried and sold as Pruteen. The dried product contains 16% nucleic acid, which is the principal reason why it is unsuitable for direct human consumption. Man's metabolism is not geared to the breakdown of dietary purines, an excess of which might lead to conditions such as gout. Animals can cope very well with high levels of nucleic acid in the diet. Poultry seem to thrive on Pruteen, and extensive toxicity tests have not shown any adverse reactions. The economic future of Pruteen depends mainly on the relative costs of the competing products, soyabean meal and fish meal. The major operating cost of Pruteen production (59%) is that of the methanol required, whereas the energy costs (for air compression, sterilization, etc.) are only 23%.

Ammonia Utilization. Two pathways for the entry of ammonia into cell metabolism exist. These are via glutamate dehydrogenase (EC 1.4.1.3) (Equation 5.1)

$$2\text{-Oxoglutarate} + NH_4^+ + NADPH \longrightarrow Glutamate + NADP^+ + H_2O \quad (5.1)$$

and the system

$$Glutamate + NH_4^+ + ATP \longrightarrow Glutamine + ADP + P_i \quad (5.2)$$

$$Glutamine + 2\text{-Oxoglutarate} + NAD(P)H + H^+ \longrightarrow$$
$$2\ Glutamate + NAD(P)^+ + H_2O \quad (5.3)$$

involving glutamine synthetase (EC 6.3.1.2) and glutamate synthase (EC 1.4.1.13) (Equations 5.2 and 5.3). Many bacterial species possess both pathways and the system which operates in the cell depends on the concentration and nature of the nitrogen source. Generally at high ammonia concentrations glutamate dehydrogenase is used. At low concentrations the glutamine synthetase/glutamate synthase pathway operates because glutamine synthetase has a much higher affinity for ammonia than does glutamate dehydrogenase (apparent K_m value < 1 mM as against 10–50 mM for glutamate dehydrogenase). This capacity to 'scavenge' for ammonia costs the cell energy in the form of 1 mol of ATP per mol ammonia assimilated (Equation 5.2). Some methylotrophic bacteria contain both of these pathways for ammonia assimilation (*Methylomonas methanica*, *Methylobacter vinelandii*, *M. bovis*) whereas others such as *Methylosinus trichosporium*, *Methylocystis methanolicus* and *Methylophilus methylotrophus* contain only the glutamine synthetase/glutamate synthase system. Under the conditions of high ammonia concentration that apply in the pressure cycle fermenter, the ATP expended in assimilating ammonia is energy wasted which could otherwise be used for the synthesis of more cell material if the glutamine synthetase/glutamate synthase system of *Methylophilus methylotrophus* could be replaced by glutamate dehydrogenase. This has recently been achieved (Windass *et al.*, 1980) using genetic engineering techniques. The glutamate dehydrogenase gene of *Escherichia coli* was introduced into the cloning vector pTB 70 and thence into a mutant of *M. methylotrophus* lacking glutamate synthase. The resulting genetically modified bacterium *M. methylotrophus* 56.2 would grow on methanol and ammonia stably without losing its 'foreign' gene and had 4–7% higher biomass yields than the wild-type organism. This represents a type of genetic improvement in an industrial organism that could not have been achieved by the traditional industrial techniques of prolonged selection, or by mutagenesis, for it represents an addition to, rather than a subtraction from, the cell's genetic information. The result is a significant improvement in the Pruteen yield and consequently improved profitability for the process.

Biotransformations using Methylotrophic Organisms

The use of microbial cells to obtain products by fermentation processes is very old. The production of alcohol for example has been known for several thousand years, and other examples of the use of industrial fermentation are of great importance, e.g. the production of citric acid and penicillin. Industrial fermentations are not necessarily anaerobic processes. The word has been used in this chapter to describe industrial processes involving the growth of micro-organisms on a large scale. This is a much looser definition than the strict Pasteurian concept of the accumulation of metabolites as a consequence of anaerobic growth of micro-organisms. With the exception of recent developments in the industrial use of alcohol in Brazil from fermentation processes (which are outside the scope of this book), most industrial fermentations are those for the production of compounds that cannot readily and cheaply be produced from petroleum, natural gas or coal. Sometimes these compounds are bulk industrial chemicals, such as citric acid, but more usually they are materials, such as drugs, that are required in kilogram rather than tonne amounts. The development of genetic manipulation techniques has opened up this area immensely, because it is now feasible to use a micro-organism whose genome has been appropriately modified to produce 'foreign' proteins, such as hormones. It is not as yet certain if these micro-organisms will grow stably in a fermenter without changing their genetic properties, and the possibility exists that it will be necessary to use them in a non-growing state, immobilized on a suitable support.

Immobilized Cells and Enzymes. A recent biotechnological development of great importance is the use of immobilized non-growing bacterial cells or immobilized purified enzymes in the industrial production of commercially important compounds. The oldest example of this technique is probably the use of beechwood shavings to immobilize the organisms used in vinegar production. The use of immobilized glucose (xylose) isomerase (EC 5.3.1.5) in the production of fructose syrups from glucose on an industrial scale is well established in the USA, but again this development is outside the scope of this book. What concerns us here is the use of methylotrophic bacteria or their enzymes in a biotechnological context. This area is not as yet established on an industrial scale, but much laboratory- and pilot-scale work is in progress, and it seems likely that commercial applications will soon be developed.

The methylotrophic enzyme system which has attracted most attention is methane mono-oxygenase (Chapter 3), which catalyses Equation 5.4

$$CH_4 + NADH + H^+ + O_2 \longrightarrow CH_3OH + NAD^+ + H_2O \qquad (5.4)$$

This enzyme, although of fundamental importance for bacterial growth on methane, may become industrially important for a quite different reason. It is a non-specific enzyme, and in addition to methane will oxidize many compounds (see Dalton, 1980), some of which cannot be transformed by traditional industrial chemical methods. The scope of this field is made even larger by the fact that there are differences of substrate specificity between the two different methane mono-oxygenase systems (Chapter 3). Transformations observed include the following types

$$R-CH_3 \longrightarrow R-CH_2OH$$

$$R-CH=CH_2 \longrightarrow R-\underset{\displaystyle \underset{O}{\diagdown\ \diagup}}{CH-CH_2}$$

and some of these might have industrial usefulness. For example, the current industrial process for converting propene (propylene) to 1,2-epoxypropane (propylene oxide) involves several different chemical steps and is rather an expensive process. The availability of a method using methane mono-oxygenase, if it could be scaled up sufficiently, would offer several advantages. When cells containing methane mono-oxygenase are exposed to propylene, they convert it into epoxypropane, which is not further metabolized or degraded. Unfortunately methane mono-oxygenase is extremely sensitive to product inhibition, and so far this problem has retarded further developments.

Unlike simple hydrolases or oxidases, mixed-function oxidases, such as methane mono-oxygenase, need reduced nicotinamide nucleotides in addition to oxygen in order to function. It is costly and impracticable to use reduced nicotinamide nucleotides with an immobilized enzyme preparation, and two alternative strategies have been proposed to cope with this problem. The first is to supply the electrons required by means of an electrochemical system. This idea is still in its infancy and the applications are still being investigated (Higgins *et al.*, 1980). The alternative approach is not to use immobilized purified enzymes, but immobilized intact bacterial cells. These can be attached to a suitable matrix (support) or entrapped within one, and when supplied with substrate will metabolize it even though they are not growing. This approach has some advantages (Drozd, 1980). Firstly it avoids the need to break the cells or purify the enzymes. Secondly, enzyme systems are almost invariably more stable in the cell than in extracts or in the purified state. Thirdly, multiple enzymic steps are possible, which is not the case with immobilized enzymes. Finally, the need for reduced nicotinamide nucleotides is met by the cell itself oxidizing the same (or another) substrate. The disadvantages are that the desired product may be metabolized further to an unwanted product, the concentration of the desired enzyme in the cell may be very low compared with an immobilized purified enzyme system, and, finally, permeability barriers may slow down or prevent penetration of the substrate into the cell. Nevertheless, the immobilized cell technique has been applied to various industrial fermentation techniques including brewing (see article by Messing, 1980). Several different methods of immobilizing microbial cells have been tried. Entrapment in polyacrylamide gel or calcium alginate are the most successful methods, but adsorption of the cells on to discs or porous brick, cellulose or substituted celluloses has also been used.

The use of methanotrophic bacteria as the active agent of an industrial hydroxylation plant would require adequate supplies of oxygen and an adequate solution to the problem of how to handle substrates that are insoluble in water. The design of a suitable fermenter might also present problems. These topics are discussed by Drozd (1980) and Dalton (1980).

Methylotrophs as Producers of Metabolites from Methanol. Japanese workers have pioneered attempts to use cultures of methylotrophic bacteria, especially those grown on methanol, as possible sources of amino acids and vitamins (Tani and Yamada, 1980). Among extracellular metabolites, strains have been isolated that accumulate glutamate, valine, leucine, serine and vitamin B_{12}, and extracellular polysaccharides, but large-scale commercial applications have not yet been reported. Among intracellular metabolites isolated in large amounts from methylotrophs are coenzyme Q_{10}, polysaccharide (glucan) and poly-β-hydroxybutyrate. The latter seems to offer possibilities as a non-petrochemical-based polymer suitable for the manufacture of plastics. This represents an area additional to single cell protein which could make use of large scale growth of methylotrophs on methanol.

Possibilities using Genetic Manipulation. If the key enzymes for methylotrophic growth using the hexulose phosphate cycle (i.e. those enzymes that are unique to methylotrophy) could be introduced into an organism that produces industrially important metabolites, e.g. citric acid from glucose, then the possibility exists of using cheap industrial methanol instead of carbohydrate as a source of such compounds. Using modern techniques of genetic manipulation it should be possible to transfer the genes for methanol dehydrogenase, hexulose phosphate synthase and hexulose phosphate isomerase into a suitable recipient organism. This may prove quite difficult, for early experiments with cloned genes from *Pseudomonas* AM1 demonstrated that it was impossible to obtain expression of such genes in *Escherichia coli* (Gautier and Bonewald, 1980).

Finally, note that the topics that have been discussed in this chapter are only indications of the wide range of exciting biotechnological developments which may be expected in this field. The student is urged to consult the references at the end of this chapter to discover further details and possibilities.

Summary

Methanogenic and methylotrophic bacteria have attracted much interest from biotechnologists. The methanogens, which have been used for nearly a century in sewage purification, offer possibilities for the conversion of excrement, whether human or from poultry and domestic animals, into combustible gas (biogas, 67% CH_4: 33% CO_2 by volume) for heating or electricity generation, and a relatively rich and almost odourless fertilizer. If applied extensively in modern intensive agriculture, this could minimize offensive odours and risks of pollution. Other possibilities under active consideration include the conversion of crop wastes, such as straw, waste from food-processing industries and even domestic refuse into biogas and odourless fertilizer. Proposals have even been made to convert crops and silage into biogas. Technical aspects are considered for those processes that are in active industrial use.

Methylotrophic bacteria and yeasts offer possibilities of obtaining several useful products as a result of microbial growth on C_1 compounds. The compounds under consideration are methane, methanol and carbon monoxide (CO) (synthesis or blast furnace gas). These are relatively cheap and abundant natural materials that offer the possibility of conversion, via only one single step involving micro-organisms, into products such as protein, amino acids, vitamin B_{12} or poly-β-hydroxybutyrate. The conversion of methanol to single-cell protein on an industrial scale has already begun. ICI are marketing Pruteen, obtained from the hexulose phosphate pathway organism *Methylophilus methylotrophus* as a replacement for fish meal or soyabean meal in poultry feedstuffs. Methane and CO represent as yet unexploited possibilities. Products other than single-cell protein have not yet been manufactured on an industrial scale.

The possible applications of enzyme systems of methanotrophic bacteria, particularly methane mono-oxygenase, are also discussed. Here again developments are only on a laboratory scale, but possibilities exist of bringing about transformations that are difficult to achieve industrially such as that of propene into 1,2-epoxypropane. The advantages and disadvantages of